AN ENGINEER'S JOURNEY IN NASA

*Space technology and policy
on the final frontier*

Edgar Zapata

Copyright © 2024 Edgar Zapata

All rights reserved

No part of this book may be reproduced, or stored in a retrieval system, or transmitted in any form or by any means, electronic, mechanical, photocopying, recording, or otherwise, without express written permission of the publisher.

An engineer's journey in NASA
Space technology and policy on the final frontier.

By Edgar Zapata

First revision 2025
Originally published in parts at zapatatalksnasa.com, 2021-2024

The views and opinions expressed are those of the author and do not reflect the official policy or position of any agency of the US government.

1. Space Policy 2. Aerospace - Technological Innovation 3. NASA 4. SpaceX 5. Space Shuttle

Front cover photograph by author, rear cover photograph, NASA

ISBN: 9798344640006

*To my wife, Arelis, always my muse
as I put my stories into words.*

CONTENTS

Title Page
Copyright
Dedication
Epigraph
Preface
*** — 1
Part 1 NASA: Lessons for today in stories from the past — 3
NASA's really declining budget — 4
Drawing the short straw — 7
The best laid schemes of NASA — 12
The call, spacesuits, and everything else — 18
NASA – TNG — 26
The unreliable narrator — 33
The Supply Chain Crisis: A Historical Perspective — 39
Planning and space exploration — 45
Unburying the lede — 51
A review of the ASAP review of NASA — 55
Two reports, two NASAs? — 58
NASA - join the club? — 62
Monty Hall, goats, the odds, and new rockets — 67
NASA, aerospace, and optimism — 73
Mind the gap — 78

Revisiting the near future of human spaceflight	83
The NASA Budget - running in place or getting ahead?	88
A NASA IG report, a story, and a question from the audience	93
Canceled X-planes, context, and NASA	99
"Rescue Party," a short story about NASA and SpaceX – written in 1946	103
NASA, space projects and context – a missing link	108
The ISS: For every beginning there is an end, or not?	114
SLS and Orion costs - the third rail of NASA cost estimating	121
* * *	127
Part 2 Reusability, sustainability, those (bothersome) "-ilities" and choices	128
Reusability, priceless.	130
Reusability - legs and fins or wings and things?	134
X-33 – The middle path?	139
You can't always get what you want, but…	144
Not in our stars	153
Contrasts	159
Life finds a way	166
About Starships, and the (not what you think) reusability we need	172
Rocketry – is it more like baking, or cooking?	176
SUSIE, space launch, and the many journeys to full reuse	181
Sustainability and NASA's human spaceflight program: We need to talk.	185
The NASA cis-lunar universe: LEO to Mars+	192
* * *	199

Part 3 Space technology, R&D, and our future exploring space	200
Technology stagnation and NASA – problem and opportunity	201
Of external tanks and Starships	204
The rise, fall and rise again of refueling – in space	210
What's old is new again – more on refueling in space	215
One word: Propellant	220
Starships mean refilling your tank, and so much more	224
Reusing, refueling, partnering - and going nuclear	231
It's a system	237
The valley of death	243
Natural and Artificial Flavors Added	248
"In the end..."	253
Is this now?	259
R&D investment and "how" – the final frontier	263
The nuts and bolts vs. NASA budgets	268
Rising wages, meet technology adoption	275
Breaking the speed of analogies	282
Of Starships and spaceplanes, and roads less traveled	290
* * *	297
Part 4 Stories of the once and wondrous Space Shuttle	298
The flow managers glossary	300
The case of the $5,000 socket	302
It ain't over till it's over	308
Make good choices	312
* * *	325
Part 5 Commercial space – and the future of NASA	326
Sustainability and space exploration	327

It's not what it looks like – the cost of ISS per year	330
Commercial space stations begin shifting the conversation to "why space"?	336
New space, a Rorschach test	342
Are you happy on average?	350
Financial risks, spaceflight, and the questions we ask	354
NASA commercial space, the 16%	358
Space benefits, stem cells, and why we're just getting started	363
A checklist for commercial space and NASA	368
NASA: Making markets, not rockets?	371
Space, playing the long game	377
Europe, ESA, the EU, and the space sector - where to next?	382
A picture worth a thousand words - flight rate, NASA and space exploration	389
Space based solar power and not losing sight of the plot	397
* * *	405
Part 6 AI, NASA and what things will come	406
I'm with the AI, and I'm here to help	407
AI, art, writing, oh - and spaceplanes	411
"I didn't understand a word you said"	417
Trust the director	423
***	427
Epilogue	429
Afterword	433
Acknowledgement	435

"It is difficult to get a man to understand something when his salary depends on his not understanding it."
-Upton Sinclair

"For a successful technology, reality must take precedence over public relations, for nature cannot be fooled."
-Richard P. Feynman

PREFACE

You are in NASA, the National Aeronautics and Space Administration – you are in the world of aerospace. We live by the big questions – how to explore our world and reach for the stars, and what will you do when you get there. For me too, why write about it? There is the temptation to belittle attempts to gather in words the ineffable experience of a rocket launch, or being a small part of preparing these spaceships for their journey, or exploring the possibilities for what's next. Shortly into their career, an engineer quickly learns they will be valued, your boss channeling Yoda, for "doing," not for "pondering." The rule is out there, an unwritten rule. With the passage of years, experiences accumulate. When sharing, there is no need to clarify these lessons, failings, and the curiosities that remain a mystery came along "in real life." The phrases arrive naturally, "We looked at that," and "That's not going to work," or "The technology is still immature." The experience follows, mentioned as if in passing, and then the question about what to do now. What will be different this time?

Writing will be required, of course, but as if interfering with the task of learning and doing. An engineer's life is incomplete without a specific sequence of events tossing words around.

The first presentation, with pretty graphs and pictures. Graphs only the author could love. Fonts that are extra-large on one slide and illegibly small on the next. Say "eyechart" and this achieves a begrudging checkmark. Done. And now, back to real work. There is a text version of what happened in the project, the technology ups and downs, the schedule, the calculations. The information vomit is barely legible. In this initial step, the aerospace sector teaches us to parse the phrases "Probably" as distinct from "More likely than not" and "Appears to." Careful with your conclusions, if there are any.

Time relaxes the topics, and one day, I discovered I might express a forceful opinion. The conference accepted my paper, where I strongly suggest the importance of a field, or a technology. Our business enjoys advocacy too much. A subjective slippery slope is part of the terrain. This phase of writing was a welcome departure from simply summarizing a project. Now the cub-reporter could finally join information with what appeared to be meaning, a lesson, a take-away, an "aha" moment. The pleasure did not last.

Our aerospace industry was more dynamic decades ago than it is given credit for if the rate of change today makes it seem things moved more slowly. Information built up faster than most people could make sense of it. Before anyone could write, they debated everything like a car crash after the fact, "what happened?" Some saw the accident. Others arrived long after but analyzed the skid marks, the direction the damaged cars faced, or the stories of the drivers to see who made sense. I count myself among both crowds.

Because we did and learned together, there were working groups, project teams, committees, and boards. Now, writing would be a merger of minds, with more value in one place than any human could ever hope to put down on paper alone. This was a new phase - the report by consensus. The most potent substance, the simplicity of valuable observations from the accumulated years of so many ingenious people, met a watering down – drip by drip, sentence by paragraph by period. The dilution of wisdom is the key characteristic of this form of writing. Our aerospace world sees reports by committee nearly every week, the academy of this, the independent auditors for that, the paper with a dozen authors.

Valuable as they are, readers are left to find the kernel of truth in the barrage of book-size reports. That is if you can make it to the end of what reads like a complicated tax form merged with the owner's manual for your dishwasher.

If you decide you will catch the headlines only, no one will blame you. There is no lack of headlines and sub-titles and key blurbs in bold. Pass by as many as you can, in case anyone asks at the bar – did you hear about that?

Some movies know this. Popularizations do wonders to explain and to inspire, to understand what all these scientists and engineers do. The producers know, there is a fascinating tidbit in all the scientific and technical garble, enough to expand and fill a script. But these audio-visual stimuli are a different beast. In the spirit of "show," don't "tell," movies about the wonders of a technology remain safely in the territory of amazement but say little about present challenges.

Writing about my experiences and what I learned and leveraging this to reflect on what's happening today in aerospace would not be a popularization of the week. When I began my blog at zapatatalksnasa.com, I intended to talk about NASA, past *and* future. But I also wanted to use my ability to "talk NASA," as if versed in an obscure language and now translating for others.

Today, news sites covering all things space and technology routinely throw around words like monopoly, duopoly, competition, composites, contracts, who's in and who's out, and how much cashed was burned like kindling. Today, that money may be private as much as public and from tax payers, if not more so. This is a good thing. If this coverage continues, it must be because there is readership, or the walk down this middle-ground between technical and accessible would have failed. This no man's land, which is not a popularization but seeks to be accessible yet informed, has been my home.

If I bring a unique flavor to this spot, it may be from the wonkish couple of thousand words. Though I prefer to think the special sauce is being frank. A reader once captured it well, though only afterward did I re-read my thoughts and realize what was happening over time. No longer with NASA, my writing did not take long to cross into the realm of "brutal honesty." If this was ever unproductive, my apologies, though the part about honesty makes me think the sin should be forgiven.

Our aerospace sector post-SpaceX (say that as if saying "post-iPhone") changed in more ways than are apparent. Now, brutal honesty is everywhere, as if the fear is gone about offending a future employer or the phone ringing with a call from the boss who received a tirade from a VP of that big, scary company. The gates are open, or better yet smashed – for now. An informed and involved public, in the business or simply enthused, is now part of the conversation. However, the part about informed begs the question about who is informing. Which brings us full circle to sharing and writing.

NASA has no lack of excellent managers. I saw these at work over about seventy-five Space Shuttle missions I was privileged to help prepare and launch. I eventually became a manager too, in a path rising through "technical" management (avoiding supervisory duties.) Knowing how to get things done within present circumstances while following a byzantine process is no small matter. To add to the challenge, circumstances, and processes will change next year. Leadership requires a different skill set.

NASA leadership has a handful of options. For management, there is resignation, doing well what Congress or the Whitehouse instructed, knowing it's nowhere close to a path to success. For leadership, you add enthusiasm. Here, too often, there is a forgetfulness about roles. NASA leadership works for the Executive branch. They work with Congress. This is not repeated often enough or understood. It's an easy and convenient slip if you believe the current administration is temporary, but your Congressional overlords will be around for decades. Adding to the logic when picking sides, projects can stretch across a generation.

If you are not resigned to poor orders, there is denial. Things will work out anyway, it turns out - be patient. After the pictures come back, or the rocks, the poor execution, the criticisms, difficulties, hearings, and near cancelations (and real ones) will be forgotten. The cheering once there will make the mess that dragged on for years before seem like so much pettiness from non-believers.

A third path is necessary. Being cheerleaders for Congressional desires or assuming time forgives all sins will not lead to success. Leadership must embrace innovation, not just in technology and things you can touch, but in getting around barriers, bureaucratic, political, or otherwise. This requires situational awareness, which much of my writing comes around to.

A spaceship. Look up and around, and you will see you are inside a spaceship. Careful where your hands or feet and elbows land. No one will care about your head should you bump it, more concerned about the damage you may have caused to that valve. Focus will make anyone forget to take in the moment. The casual – I am not a tourist – vibe comes along, and we believe we have already been here six months and should stop saying "Wow." This was me. Though not always. This focus makes it difficult to reflect on what happened at the moment and long after. Yet, it is the responsibility of everyone in our space sector to share and to do so in a way that is accessible. My attempt here is lacking, as we said in those obtuse papers – more likely than not. But the bits of knowledge, stories, experiences, and outlooks we share might do wonders when pieced together.

Far removed from technology and gadgets, our aerospace sector has a doppelganger called policy. If the stuff we can touch and hold carries us into endless conversations about "what," we forget at our own risk, there is another world all about "why." These two cousins have a common ancestor, decisions about what comes next. One day, that decision is a spacecraft or an engine, but also a contract, a position NASA or our DOD brethren find themselves in for another generation. Suddenly, we see the writing on the wall. How did this happen? How did we let this happen? Wouldn't it have been better to have competition and a vibrant aerospace sector that makes us look back every five years, like with our cell phones or cars, at how antiquated the last model was?

In this spirit, I have gathered my writing here, distilling it to include the blogs I feel took my experience and made sense of it. Writing will do that. Write about something you hold near and dear in your lived experience, formative even. Does the writing lead to an ending like water flowing downhill? Or is it difficult to make it past an invisible wall, instead devolving into a mere blow by blow of what happened? Is it Wiki-like information, dates, times, and what someone said? Perhaps this was the spirit of the saying attributed to the pioneering physicist Ernest Rutherford. If you can't explain something complex to the barmaid, maybe you don't really understand it. I'll extend that – if we as a scientific and engineering aerospace community can't explain our experiences simply to a neighbor or family, perhaps we don't understand what they meant.

As with the second law of thermodynamics, entropy always wins. Hyperlinks will break, or some perfect reference that seems current to me now will make no sense to readers tomorrow. Previous lessons will be overcome by events. This was a lesson too. The chances a complex, ambitious project succeeds diminish rapidly over time as time keeps throwing the twelve-sided die. On one of those throws, the project will lose. Act with urgency. Be impatient. Or later, realize the project failing was inevitable due to moving too slowly.

It was a sunny, breezy day at the launch pad. I am literally at the top of the Space Shuttle – the Orbiter, External Tank, and Solid Rockets are all below, pointing up. The operation today is minor, nothing to write home about. The technicians climb up through a network of pipes, crawling to get atop the liquid oxygen vent hood or "beanie cap," as we called it. More likely thinking of it as a joke, how obviously I won't take up the offer, I hear "You can climb up, just grab a safety harness." In a moment, I strap on the belt, hooking it here then there, and I am atop the beanie cap. I would soon move on to another Shuttle system, so perhaps I was ready to take in this moment. The techs had the job in hand.

At the launch pad, an engineer quickly finds out if they are afraid of heights. I learned I was not.

The guardrail on the small circular platform is precariously less than waist-high. Though the height creates the illusion the bar is only knee high. Still, I can't resist but look around, the waves ahead coming ashore, the clear sky, the miles of land leading back to the vehicle assembly building, the launch pad circumference. I lean slightly over the railing, looking down on the nose of the Shuttle – from nearly straight above. Perhaps one day, I'll make sense of this. Today, there are no words.

Edgar Zapata
September 30, 2024

* * *

Part I

PART 1 NASA: LESSONS FOR TODAY IN STORIES FROM THE PAST

NASA is an endless topic. There is technology, rockets, propulsion, politics, promises, tragedies, current controversies, and dreams of what's to come. We will always dream. The background for this is lost in a stampede of stories with the luxury of a marvelous picture or a video and a lift-off, a cheering crowd at the start-up – with the right music, even better. Essential context receives benign neglect. We know, "No bucks, no Buck Rogers." But money and budgets lie on the surface of a complex system of people and processes, expectations, and planning (or lack thereof). The challenge is asking the right questions to understand this system and its connection to what's in the headlines. We must if we are to fix what's broken today and move on to new, more exciting problems tomorrow.

NASA'S REALLY DECLINING BUDGET

It was the late 90s, and everyone was so happy to hear the budget would remain flat because, after all, flat was the new up. A lot has happened since. Glancing at NASA's recent budgets seem to show good times since 2014, with funding going up generously every year. During this time, NASA's budget grew by 29%, just under the federal budget growth in the same period. Except that's not the whole story.

How do NASA dollars look if we consider inflation? This would be a good time to say, "I don't know."

If we take the official NASA Inflation Index, NASA has less purchasing power today than in 1995, 19% less. The calculation's simple, and the result doesn't sound too bad. After all, we're fortunate to live in a time when we're seeing launch costs drop in good part due to NASA's partnerships for commercial cargo to the ISS, an investment that has paid off handsomely. There are also new partnerships to develop landers on a commercial basis, not old-style "cost-plus" contracts paying to cut grass by the hour (guess how long that takes). This must all balance out, right? Some deflation here, some inflation there, all captured in the index and the result. Not quite. An average might be correct, but not always useful. When you have one foot in hot water and one foot in cold water, it does not add up to comfortable. Here this is especially so, as "commercial" dollars (at the bottom of the graph) are small parts of NASA's huge spaceflight portfolio.

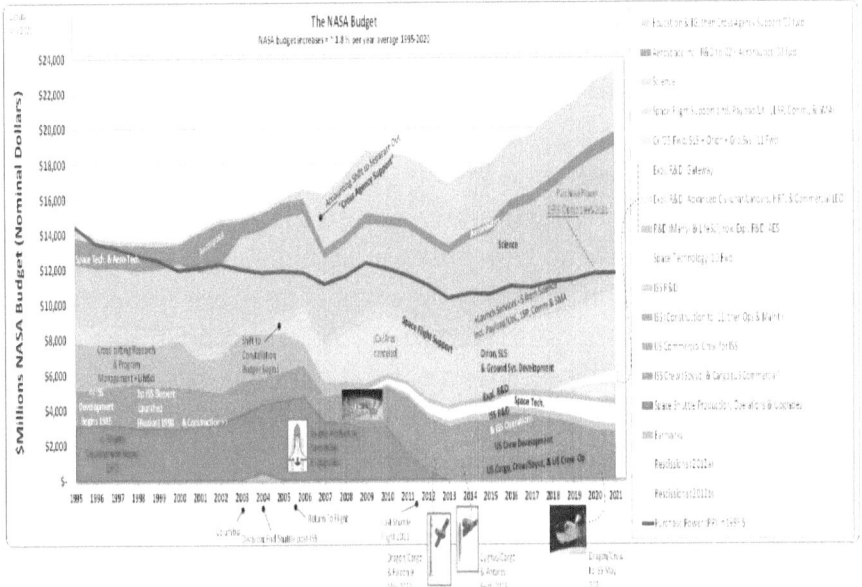

NASA's yearly budget since 1995. What appears at first glance to have gone up over time has instead gone down, as shown with the blue line, NASA's declining purchasing power.

Then there is more uncertainty. If we ask what the effect of inflation is on NASA's budget, we have to ask how close inflation indexes may be to the mark. Great care goes into these inflation rates, but even the best of intentions have limits. You won't find rocket engines, spacecraft, and flight-weight composite tanks next to butter, steak, or healthcare in any general inflation index that informs the NASA index. Even adjusting for this, who really believes what NASA buys as goods and services have tracked along with much higher volume aerospace goods, electronics, or chips versus ever more expensive healthcare or education? There are excellent arguments that real-world inflation in the wild runs much higher than any official inflation reporting will ever reveal. The agency budget guidance sent along with the indexes was always fixated on what more the government might spend next year, not on fixing an outcome and calculating the updated dollars needed. Agencies are told from on high what inflation will be. Feedback is not requested.

All to say, if we don't like losing a fifth of NASA's purchasing power over the last decades, maybe we don't know just how much purchase power has really been lost. As likely, we will find a better answer even less appealing. This naturally leads to the need for innovation and competition and back to the "commercial" lines at the bottom of the budget graph. It's entirely possible the only way to take all the more NASA dollars every year (as long as that lasts) and actually make them behave as more dollars instead of really being less is to disrupt, innovate, and change more parts of the NASA budget just as those commercial lines did.

The alternative? Well, we could do the math, take out the NASA Inflation Index, and find some comfort, looking back only so far, declaring we have turned a curve. Notably, the generosity in NASA's budget since 2014 was a smidgen behind the growth in the US Federal budget. The federal budget windfall since 2014 didn't single out NASA as uniquely important. NASA has just been on the same ride as everyone else.

Better yet, we could look at those lines at the bottom of the budget graph and see a future where only a bit more or even fewer dollars can still buy more value when (not if) that day comes.

DRAWING THE SHORT STRAW

"I think it's fair to say that our review group drew the short straw, and I drew the shortest by having to actually do this presentation."

-DR. SALLY RIDE, 2009

Dr. Sally Ride at the 2009 Review of Human Space Flight Plans Committee. NASA.

It is August 2009, and Sally Ride is about to present charts about NASA's possible future. This was a meeting of the White House chartered "Review of Human Spaceflight Plans Committee," not the first team (and probably not the last) to take a hard look at the future of space exploration and NASA. In 1988, a few weeks after I first arrived at Kennedy Space Center, I was given a book by my new supervisor -

"Pioneering the Space Frontier." This went on the reading pile alongside endless training material and Space Shuttle procedures to learn inside and out. The book was from a commission formed by Congress the year before the loss of Challenger in 1986. Eventually, reading it as part of drinking from a fire hose of information at the time, I was awed by everything that might be around the corner in the next 50 years. It struck me that a good part of this fantastic vision of the future could fall within my years at NASA. With good health, it could fall completely within my lifetime. There was a lot to do. Better get started, I thought.

"Pioneering the Space Frontier," 1986. NASA. Artwork: Robert McCall.

Twenty-one years after first reading about what could be ahead for space exploration, I would support a similar and grand committee. The questions had not changed for NASA as for anyone, where are you going, and what will you do when you get there? The "Review of Human Spaceflight Plans Committee" worked hard on these questions. Sally Ride began her presentation - *"I think it's fair to say that our review group drew the short straw, and I drew the shortest by having to actually do this presentation,"* she said. To say "drew the short straw" was right on the mark. This was not where you got lucky and won. Physics, rockets, and technology are the bread and butter of engineers and scientists. These are comfortable topics, defined and often exciting. At NASA, I saw very different reactions to the other important questions - what might it all cost? Does it add up? How soon? These questions made people uncomfortable. (Parenthetically, I answered questions like these for a few years before the 2009 committee. When the committee came about, I would try to answer similar questions for NASA Headquarters as part of the "Committee Staff." I already appreciated the challenge of answering questions about dollars, technology, and NASA for a broad audience that did not have secret decoder rings. This continues to be a worthwhile task. Or trying at least, and sometimes succeeding.)

As Sally Ride ended her presentation about the multi-verse of possible NASA budgets vs. possible projects, it would have been impossible to see a singular future ahead. This was the last time to date that there would be any public presentation of this sort. The infamous "sand charts" have not been seen since. They are now near extinct, or at least on the endangered species list.

And yet, here we are in 2021, with many possibilities for NASA to invest in partnerships, commercial services, and new projects galore. Where are the sand charts looking at these possibilities? Might they still be spotted somewhere in the wild? Saying the team in 2009 drew the short straw begins to answer this question. No sand charts add up to a compelling and beautiful vision of what may be as in those earlier visions from 1986. This is because the sand charts fail to add up at all, even for just what they are – the dull press of numbers against reality. They could hardly be blamed for being obscure and uninspiring. But failing to add up is unpardonable. At least if adding up, the inspiring pictures of what may be could follow and the numbers could just as well be skipped over. However the numbers didn't add up then for those next steps, enlisting the artists and our imaginations. And they still don't.

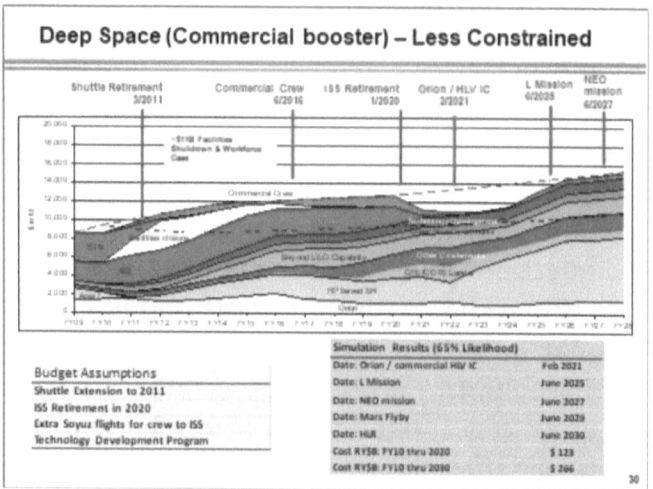

One of many "sand charts" presented by Dr. Sally Ride in 2009.

* * *

There is no spoon.

My last public foray into sand charts would be in a paper I published in 2017. There, I covered plenty of possible futures that might lie ahead for NASA's multi-verse of sorts. Yet what might be done for some stretch of dollars, technology, and "what-ifs" still draws the short straw. There is not the same level of awe-inspiring pictures to be read into or from a series of sand charts, *at least in so far as these are thought of as just NASA projects.* And therein lies the rub, pointing the way out. The dilemma was always the same, a grand, beautiful vision of what may come, the artist capturing the spirit of progress and endless possibility – courtesy of NASA. Endless possibility and NASA budgets that are not endless do not attend the same party. The answer is in realizing, as the boy said in The Matrix, "There is no spoon." Try to bend budgets to fit projects and dreams, and you soon find the budgets just don't bend. This points to the way out, that there are no budgets, there are no costs - but there are investments.

Off the beaten path, there is a way forward. There was never a way to have an Inspiration4 mission with four civilians shown on a NASA budget sand chart. Or a hundred of these. One day, we may say the same for an assortment of private space stations. Or for companies that provide refilling services for stages in space. Similarly, when viewed as layers of NASA investment dollars, daring visions can become reality, except they are on someone else's sand charts, not NASA's. If the public and the private join as successfully as before, it won't feel like anyone drew the short straw.

THE BEST LAID SCHEMES OF NASA

But Mousie, thou art no thy-lane, In proving foresight may be vain:
The best laid schemes o' Mice an' Men
    Gang aft agley,
An' lea'e us nought but grief an' pain,
    For promis'd joy!

-ROBERT BURNS

ON TURNING HER UP IN HER NEST, WITH THE PLOUGH,
NOVEMBER 1785

This was not my first rodeo. Word had spread that I was familiar with strategic planning. It had been a while, but the predictable invite to a kickoff meeting came, and I accepted. This was sometime after the end of the NASA Constellation return to the Moon program when the path immediately ahead changed, but the long-term direction mostly remained on its original course. There would no longer be an Ares I "stick" rocket with a crew capsule placed atop a Shuttle solid rocket booster. The near-term date for the end of the International Space Station was also gone from future plans. Practically, and in the near term, the cancellation of the Constellation program was merely the end of one project to develop a small stage and engine that went atop the Ares I Shuttle-derived solid rocket booster. Even so, the engine part of that program continued a few years longer. As was the norm when these shifts happened, plans were dusted off. Teams would review and refresh what had come before. This was one of those teams.

Collect the whole set! NASA vision documents, strategic plans, technology plans, spaceflight plans, center plans, master plans, and implementation plans, circa 1986 - 2005.

I first became involved in long-term planning at NASA when the future was Space Shuttle flights as far as the eye could see. It was the early 1990s, the Space Station was coming (though not yet international), and the Shuttle manifest was booked years out. It would be a while before the realization hit that the Shuttle manifest, the schedule for what launches when, was filled with space station payloads unlikely to show up as planned. Right now, if the lack of a "vision thing" was a flaw, this was just something more to fix. The circumstances at the time allowed some heady plans and roadmaps, with a dissection of the differences between why, what, and how. Not being binding, plans had a latitude that could only come from assuming they were merely fodder for stirring a stimulating picture of tomorrow. Confirming this, early planning teams were filled with post-Challenger hires. In retrospect, this was a sign the assignment was seen as harmless.

Eventually, there were presentations. A place is not a plan. A place is a destination. Going to the Moon or Mars was an objective. Strategy is something else entirely. In all cases, a strategy was not a particular rocket. That was technology. Finally, this all required asking "why" many times, purpose and mission at the top of the edifice.

The passage of time was not kind to these plans. I was eventually on more teams drawing up NASA plans or roadmaps, labeled one or the other according to how specific. All these plans saw ample review. The teams collected comments and made comments to the comments (with little modification after). After the cycle was complete came the awards (I received a few), followed by glossy, magazine-style books as decor for the coffee tables in all the front offices. I never could come to discard these books, placing each one on my shelf alongside all the others.

* * *

The trouble with plans.

Plans are like the tribbles from Star Trek. Initially comforting, but trouble as they procreate. As a rule, attempts to improve plans that came before meet a resistance proportional to the age of the plan. From experience and too much optimism, a small cadre at Kennedy Space Center grew quite familiar with all these variations in planning and the rules. That experience was not always effective, every current purveyor of a plan having much less latitude than the prior.

Experience said to begin with the common tactic of converting a plan into plain English with why, what, and how versus the word salad only large, complex organizations think is readable. Yet this tack failed surprisingly often. From this failure, something wanted became something planned, or a means to the end came to be called an objective on a par with the end itself. Concrete sets strong and only gets stronger over time.

> "Plans are worthless, but planning is everything."
>
> -PRESIDENT DWIGHT D. EISENHOWER

Miraculously, although perhaps predictably, all these NASA plans did create value – in planning. Dwight D. Eisenhower once said, *"Plans are worthless, but planning is everything."* Process can be belittled, especially in a start-up silicon-valley "just do it" mindset. Patience for plans can be short when it seems planning more means doing less. As hard as it may be to admit though, if there were ever some forceful thinking about the future, arguments and raised voices, and a consensus at the end, you were just in a planning meeting. As with writing, engaging the whole brain with little forgiving of being vague, or lacking a clear start and end, forces along difficult questions. Even the shortcomings of plans and what is not written survive in the experiences of those who were planning.

Recently, NASA leadership offered a new vision for humans and space exploration. Things to do, what may come as a result, and why were all rolled up into an elegant handful of lines from the head of NASA Human Spaceflight. Exploring the solar system is not a small number of thoughts to reduce, leaving just the cream. As usual, there is a mash-up of purpose ("return benefit to Earth"), a means to an end ("commercial human spaceflight," "coalitions of partnerships," "sustainably"), goals ("deliver more"), and why ("new discoveries," "knowledge").

> *NASA HEOMD (Human Exploration and Operations Mission Directorate) Vision for Space Exploration*
>
> *Make commercial human spaceflight to low-Earth orbit a robust, sustainable enterprise with many providers and a wide range of private and public users*
>
> *Build a coalition of partnerships with industry, nations and academia that will help us send astronauts to the Moon quickly and sustainably, together*
>
> *Deliver more missions, more science, more technology, and more innovation at a better value to the American taxpayer*
>
> *Make new discoveries, expand human knowledge, and push human presence deeper into the solar system*
>
> *Return benefit to Earth*
>
> *-NASA*

As to that meeting on plans I found myself in at Kennedy not too many years ago, that would turn out to be for a review and a cleanup for plans already out there. Nothing new was expected. The times called for a refresh, not re-planning. Perhaps another day. Seeing nothing productive in the task,

I managed to steer clear of the job with a handy excuse, too many balls in the air at the time and all that.

Since then, still more plans have come and gone. In all this, there is a natural temptation to mull about the best-laid schemes of mice and men, as if we are the mouse whose nest has been turned up in Burns' poem. A nest turned up, over and over. The farmer recognizes a kinship, plans, and life are as quickly turned up for him. Perhaps the mouse and the farmer should have seen NASA's plans.

Right along, there comes the temptation to read a lot into what's in today's space exploration vision/plan. Hanging one's hat on a particular word, or even better yet, a whole phrase from such pronouncements, is a bureaucratic art form. In the end, the plan will certainly change. However, as with previous NASA visions, strategies, and plans, a finished plan may not be the point.

THE CALL, SPACESUITS, AND EVERYTHING ELSE

"The gift shop is down that way, toward the lobby, past the spacesuit." These are not the directions anyone can give a visitor lost among a run of offices and cubicles. You get away with this in the Kennedy Space Center Neil Armstrong Operations and Checkout building, home today to the Orion spacecraft, home once upon a time to Apollo hardware – including spacesuits. If we want to go to the Moon again and go out for a walk, however we get there, we will need spacesuits. That there is one in the lobby of our building under bright halogen lights, behind glass, would seem to say that new spacesuits are just a matter of looking up the sewing pattern and selecting a size.

If only it were that simple. Better yet, imagine a dramatic moment – the urgency, the mission, the last-minute secondary character asking, "Where are the space suits?' The blank faces as everyone realizes there are none. Followed by the rush to unseal the glass case with the spacesuit in the lobby. The quick cut to the sound of sewing machines and people reverse engineering a helmet and the zillion parts of a portable life support system. A good director would keep these 42 seconds in the movie to lighten things up a notch.

This is not that movie. Reality is more like the joke by

industrialist J. Paul Getty. If you owe the bank 100 dollars, and you can't pay it back, you have a problem. If you owe the bank 100 million dollars, and you can't pay it back, then the bank has a problem. It is now public knowledge that NASA Spacesuits might cost a billion dollars when all is said and done. Billion with a B. Yet this comes after a request for industry ideas with NASA considering partnerships for spacesuits as a service – much like other partnerships, paying for the ride but not owning the car. Does NASA have a problem, or does the US aerospace industry? More constructively, does everyone have an opportunity?

A spacesuit in the lobby of the Kennedy Space Center Operations and Checkout building.

We might want a spacesuit to go to the supermarket during these peculiar times. We would look at the reviews on our favorite shopping site, avoiding the spacesuit with leaks. Water in the helmet obstructing your beautiful view of the Earth, not acceptable. Not to mention life-threatening. Size is also an issue, and since sometimes the returns are not free, we would pay special attention to having a suit with just the right fit, especially if you run small. Needing a spacesuit and not having your size in stock is also unacceptable.

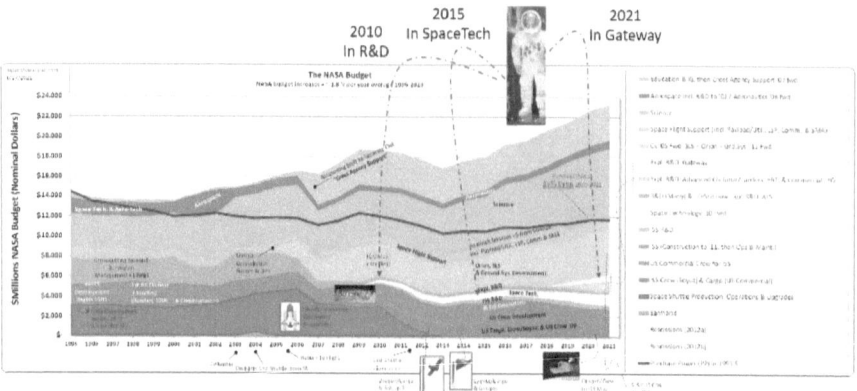

Where were those hundreds of millions that might one day add up to a billion? There's a bit here and a bit there if you go looking for spacesuit dollars.

In this real yet peculiar world, with an urgent need for spacesuits of the kind for walking on the Moon (not the zero-gravity space station kind, which is different), which glass needs breaking? And does the "in case of emergency" hammer cost a billion dollars? The answer for the curious is likely not the dramatic movie moment or as viscerally satisfying.

Instead, the movie scene *is based on true events, with names left out to protect the guilty, and* begins with a phone call. The topic is a costly project, also unnamed. For all intents, the project could also be spacesuits. Caller #1 knows the project well and is a well-respected analyst who works closely with the project's engineers. This person has boots on the ground. Caller #2 is purposefully at an arm's length from the project, independent. For a review team checking if things add up, being at arm's length is an advantage. For our characters, think "someone who did the taxes" and "someone who sees if the numbers smell right."

"I see your numbers. I just don't understand them," caller #1 starts, with a voice that says they have had a long day, and we are skipping the usual chit-chat and catching up and how are the kids. Today, we go straight to the point.

"Well, that number is the grand total. It's the sum from all the sources in the notes," says caller #2, the independent reviewer. He is going for that help-line kind of voice. Sensing something is askew with his caller, he proceeds with caution.

"That's impossible. You just don't understand." Yes, the tone has definitely shifted. This is not about figuring out something together, so more meetings and checking calendars. Is it leaning to let's get it over with?

The independent analyst, always wondering if the last edits among the never-ending edits propagated an error through every spreadsheet in the galaxy (they are all connected after all), jumps to safety with a question. "Are our definitions different?" The number looks good, he thinks, after all – at most, an error in that last edit might throw off a decimal.

"No, it's just not costs, you see." The call shifts again; will this be a short call?

Analysts have patience in abundance for everyone except other analysts. The reviewer pauses - there is a possibility, a trace of what the problem is. But this topic has not come up in years. Could it be that infamous notion rearing its head again?

"I think perhaps we are talking about different costs, the spacesuit itself versus *everything else*?" says the independent reviewer.

"Well, everything else is not *really* part of the cost." Bingo. We have found the cause of our failure to communicate. The "*really*" is pronounced stretching the "ee," where I (you figured out who the independent analyst was, right?) could almost see my caller make air quotes with their hands.

This brings us to *everything else*. Adding up budgets, wholly spent with rare exceptions, should be what something "costs" NASA. This simple view is not so simple in everyone's eyes.

Sometimes, the budget says one thing, and the funds are spent on something creatively close enough. Research and development, ill-defined as it often is, is a poster child for this. Alternately, Congress provides specific funds for specific projects, and *everything* related to that project is paid from those funds (and sometimes even more taken from elsewhere).

The failure to communicate on this call was not the first problem. It was mainly the second. Consider the NASA Space Launch System, the Orion spacecraft, and the Kennedy Space Center ground operations projects. After that, a curious mind might subtract the awards to the Lockheeds and Boeings and make a pie chart because why not. Then, there is a little more math, as other numbers are known, like ground operations. Surprisingly, but not to analysts, there is still a sizable remainder.

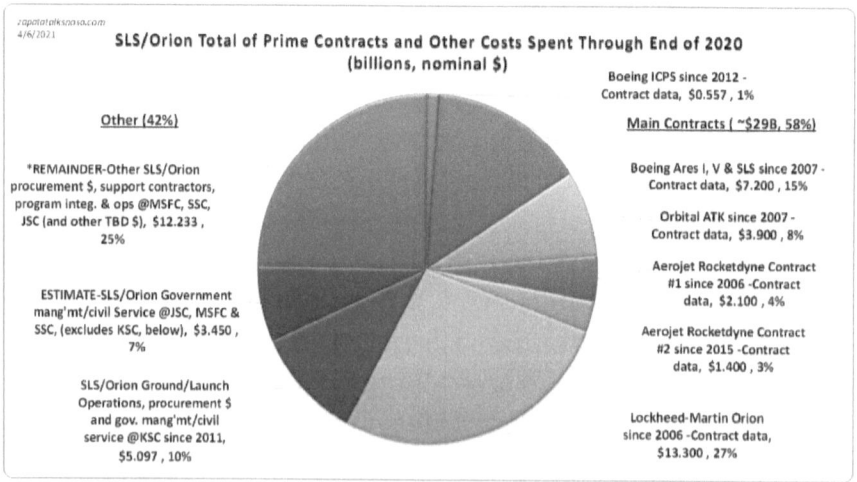

In NASA's "cost-plus" projects, there are the headline-grabbing "prime" contract awards, mainly on the right side of this pie. Then there is everything else, just as much a project cost, just as necessary, but flying under the radar.

Of course, the totals will be high - they come from adding up huge budgets year after year. That is unless you believe – that is not what that *reeealy* costs. Why not even just the big awards are what that project *reeealy* costs. All of this is about the infamous notion – first seen in the wild in the mid-1990s, whereby the Space Shuttle really cost very little because if it did not exist, NASA would still pay to hold on to and keep *everything else*. (My Department of Defense brethren said the way NASA accounted for dollars was ripe for making projects an easy target of criticism. They gave themselves credit for having figured out how to turn large numbers into endless smaller numbers. Divide and you won't be conquered.)

The Kennedy Space Center Vehicle Assembly Building (VAB), 2019.

I have come to call this the "VAB Effect." Albeit, this is really about people and organizations and not just buildings. If you only have one customer, inevitably, *everything else*, buildings and people, really adds up fast and *really* is part of the cost. (This is the part where I could cause eye-balls to roll upwards past what would seem physically possible. It would be a talk about indirect costs vs. overhead, institutional vs. corporate, knowledge vs. capability, support vs. prime, federal employee vs. contractor, and enough inside baseball to create new research avenues to help people with trouble falling asleep.) Imagine a restaurant keeping a large kitchen staff on salary, minus a menu, a specific meal request, or a customer. For years. Then saying, "You see, the food here hardly costs anything!"

As that call ended, we agreed to disagree. The most important question will not be if a number is a billion here or there, or twenty-three, or more when adding it all under one lunar moment. It is not even a question of bookkeeping or keeping some standing capability needed for space exploration - the burner on warm, but not cooking.

Inevitably, history decides on the answer by adding up all the budgets as costs, as with Apollo bookkeeping. This is the logic of grouping results vs. dollars. (The independent analyst nailed it.)

The important question is about doing what might cost a billion, with innovation and new thinking, for much less. NASA requesting ideas is one thing. Taking up an offer is quite another. Which glass case needs breaking into? With urgency. The question that will follow 2.8 minutes later will be what to do with the VAB and everything else?

Building the Future Spacesuit: "Thanks to some funding from the NASA Institute for Advanced Concepts, we were able to gather a team to begin the practical work that would test our hypothesis." NASA.

NASA – TNG

The award said it was amazing what I could accomplish, "with an endless supply of NASA interns." This much was true, as they did excellent work. I mentored many students in their summers at the Kennedy Space Center. They always amazed me with a refreshing perspective on what might otherwise be a daily grind of spreadsheets, models and meetings. This was especially so with interns that possessed what I thought proper qualifications to work at NASA - working well in a team, being easily entertained, and having a sense of humor. This year's intern added a memorable phrase to a recognition award for me, her mentor. A sense of humor, check!

That summer, I also discovered enough post-it notes were in stock to cover every square inch of my office. Including every object on my desk. And the desk. Perhaps my post-it notes with to-dos for my intern had gotten out of hand? I made a mental note that the prank involved help from other interns. Works well in a team, check!

I'm recently retired, but once upon a time I was a new hire on the other end of this equation. It was me hoping to show the proper qualifications to work at NASA (easily entertained, check!)

When I arrived, NASA had as many people my age, 25 to 29, as there were 55 to 59. The NASA hiring spree after the loss of Challenger was well set to fill the ranks of the soon to retire Apollo generation. By the late 1990s it seemed there was free food at least once a week, just from all the going away parties as the 1960s hires retired.

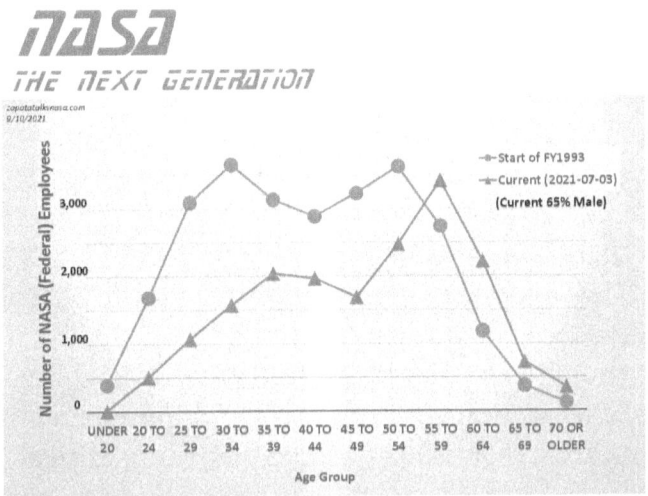

The distribution of NASA employees by age in 1993 versus 2021. Data: NASA.

The Apollo generation built multiple spaceflight programs, *from scratch*. They went from rockets that did not exist to massive Saturn rockets, then to the Moon enough times for the networks to decide it was too boring to interrupt the regularly scheduled programming. In the early 1970s this generation moved to the nascent Shuttle program, making a spaceplane a reality. Eventually, this would launch enough times that it was also thought so routine. If you stayed around long enough, and moved around as well, you were part of putting a man on the Moon, creating a reusable Shuttle, then a space station, and starting a program to return to the Moon (the canceled Constellation program.) Though I knew of only a handful at Kennedy who could say this.

You would be part of a scarce group if you arrived after Apollo, but in time to be part of the Shuttle development. More likely, and the larger generation, were those who began with the Shuttle already flying. My sizable post-Challenger generation was in that 25 to 29 bracket in 1993. This generation would see the Shuttle as a given, operational, and a space station program struggling to figure out its next steps. The Space Launch System/Orion followed, and so too an earlier, less advertised program with a new way for NASA to get its cargo and crew to orbit.

Jumping to the present, a lopsidedness sticks out. There are no longer as many people in NASA age 25 to 29 as 55 to 59. It's easy but premature to dismiss this by simply pointing out the total number of NASA employees has dropped. The temptation is to say there were 26,146 NASA employees and 1993, versus 18,084 today, and today's curve will soon flatten out too, just at a lower level.

This thought is comforting but incorrect. And as usually happens when exploring data, you see an immediate problem, then think about something hidden and more critical.

 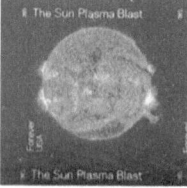

Then and now. A first-class stamp was 29 cents in 1992. It is 58 cents today. One thing has not changed–NASA is always been a popular pick for the US postal service.

The immediate problem is the band 25 to 29 today should be higher to match the group over 40 and flatten out the curve even as the number of employees drops. This can be fixed with a hiring spree to match any upcoming retirements. I was often told by those I left behind, "I will be leaving after the first SLS launch." Predicting what will really happen is a personnel science, so fixing this may be manageable.

When looking at age groups, then and now, what is critical is asking if anything has changed. Has enough changed to make the lines apples and oranges? It's been a few decades, and a lot has changed.

Fridays, and as the opportunity arose, I would get the interns out and about to see real hardware. There's nothing like a Shuttle orbiter. It just does not get old. If you crawl around enough to ruin your jeans, you might just learn something new. And who knows, ripped jeans might be a thing one day. Or it could just be a beautiful day to go to the launch pad. This was a time when a NASA badge got you anywhere. Just act like you own the place I said, which in a sense, you do.

Technicians were always glad to talk to "the young people." These field trips were necessary to see and appreciate the real world behind the mind-numbing amount of information we asked our students to absorb back at their desks. Kennedy educational programs went further, arranging even more lectures, with visits to facilities and flight hardware.

It's a cliché that NASA spends 90 cents of every dollar it receives from Congress. The less traveled version is that the 10 cents are the NASA-badged federal employee workforce. The first version usually gets around more. These are "personnel" in NASA-speak, and today the number is about 12 cents of every dollar (twenty years ago, it was 15 cents).

It's also somewhat more complicated than a simple number, but close enough. You can jump from 10 cents on the dollar to answering the question – how much of any project's costs are NASA employees? The answer is, not surprisingly, usually around 10 percent. That is until a new, less advertised program with an entirely new way to get NASA to low Earth orbit after the Shuttle program's end. The commercial cargo program for the ISS ran at *3 percent*. Playing a little 3-card Monty, where is the other 7 percent?

> *Of the $500 million originally allocated in 2006, C3PO designated only 3 percent for program management, leaving 97 percent, or about $485 million, to give directly to the commercial partners.*
>
> NASA COMMERCIAL ORBITAL TRANSPORTATION
> SERVICES FINAL REPORT

An essential ingredient of commercial programs is about property. Who's is it? (And can you walk in like you own the place? No.) The Space Shuttle, as with the SLS or Orion today, was government property, as were the facilities and every nut, bolt and tile. Hardware arrived at Kennedy and NASA employees oversaw their contractor counterparts to prepare it all for launch. As we were fond of saying, "Kennedy is where it all comes together."

In commercial programs, the hardware is the property of the private sector partner (say SpaceX, Northrop Grumman, or Boeing.) Their facilities, though on government property, are not government-owned either. These may be leased from the government, but more likely, they are wholly owned by the partner (just on leased land).

So, a NASA badge will not get you around as much today as when I walked about anywhere or when I hollered "field trip" to the interns. And therein lies the rub. There is a hidden challenge in the data. How do we make sure NASA – the Next Generation grasps (literally) the space system and people they will manage, as real, palpable, in front of them, in their hands? The call at one in the morning about some problem? "I will be right in," I said, rather than debate if I understood what I am being asked to approve over the phone. This happened many times in my career. After all, the hardware was just a walk away or just a short drive away in the middle of the night. And yes, bunny suit in hand, or safety harness, or hardhat, I'm going to see it myself, up close. Then we can talk.

If you are of a school that you have to get your hands dirty to learn, if you are to one day run the place and do even better than when you showed up, then NASA – *we have a problem*. These days of Zoom and Teams and remote work add even more distance from the reality of stuck valves, a misaligned jig, or a firing room console. But that exists atop more remoteness, the kind that comes with going commercial.

Feeling a little dated? These were our launch room computers circa the late 1990s. The modern computer on the far left was a recent addition back then, used for monitoring nearly everything on the Shuttle, but it could not send commands. NASA.

The Kennedy commercial crew office is a case in point. Soon after commercial cargo to the ISS established a sea-change inside NASA, a similar approach came to getting crew to the ISS. The commercial crew office at Kennedy numbers about fifty NASA employees. Other centers contribute more personnel, but not much more. The simple math is a program with the yearly dollars of the commercial crew program would traditionally have hundreds more NASA employees. Consistent with a commercial mindset focused on results and less on activity and encouraging innovation, government management is leaner. There are fewer cooks in the kitchen, and this is good. Responsibilities are well defined. *Nonetheless, everyone who will make critical decisions as a NASA employee must get quality time in some kitchen.*

It's easy to overlook how technology and innovation go hand in hand with organizational change and disruption. The smaller, leaner commercial program offices move NASA forward, but with after-shocks. You have to have once been close to the hardware, just a walk away, to later be able to step back. But such a trend is not sustainable if no one ever had the close-in, hands-on experience one day.

Oddly, being leaner, to date, a small minority of NASA employees are seeing these new ways of doing business first hand. But having much less NASA in one place, meaning much more in another, can only last so long as NASA shifts increasingly toward commercial programs. This is an excellent new problem to have as a consequence of fixing another. It seems there is no lack of adventures for the next generation. Or to-dos. Let's put that on a post-it.

THE UNRELIABLE NARRATOR

There is the movie "The Sixth Sense," and there are NASA cost estimates. I am not sure which is a better example of the unreliable narrator. "Call me Ishmael." Why should I call you Ishmael? Most people would say, "My name is Ishmael." What are you hiding? Of course, the first version sounds more intriguing. I want to know more. It turns out it's a whale of a tale, and the whale gets bigger with every telling.

Having been on one too many NASA cost estimating teams, I came to realize the critical question was *not* "what will it cost?" The real intriguing question was, "does anyone really want to know what it will cost?" I was told, many times frankly, if we said what it might cost, we'd never get any projects approved in NASA ever again. One time the team leader pulled their chair toward mine, knee to knee, to convey this, after perusing my analysis. Awkward, but ineffective. Initially, and erroneously, I thought I wanted to nail the number so years later, I could whip out my estimate when the project finished. I would go to the closing paragraphs and the final number, always with a give or take. It would be close to the mark, close to how it really turned out. Not the case.

I actually did this once. I learned rule #1 in cost estimating – no one ever, at all, under any circumstances, likes to hear "I told you so." This much is constant, even in alternate universes. I had been told cost estimating was a learning process, and this was true. Though to be truthful, I pulled off the correct cost estimate by cheating. Lacking a time travel machine then, I had discovered that a great starting point to the question "how much will it cost?" was "how much do you have?". This cheat does not always work, and I discovered a better method was to couple "how much do you have? to "how long do you have?" Time is money after all.

It's worth going back to the beginning. I had just arrived at Kennedy, noticing on the third day at the office that a large half-inch thick document was left in front of the locked doors every morning. Stacks of these were here and there as well for the taking (if you were first, you fought with the string tying them all together). Like an early morning newspaper on everyone's stoop, these were everywhere around the space center, from Headquarters to our office complexes to the gargantuan VAB. I asked, "can I keep one?" Yes, of course. There were many more to come. Even a smaller one with just the highlights most days. At the time, "save a tree" was not yet a thing.

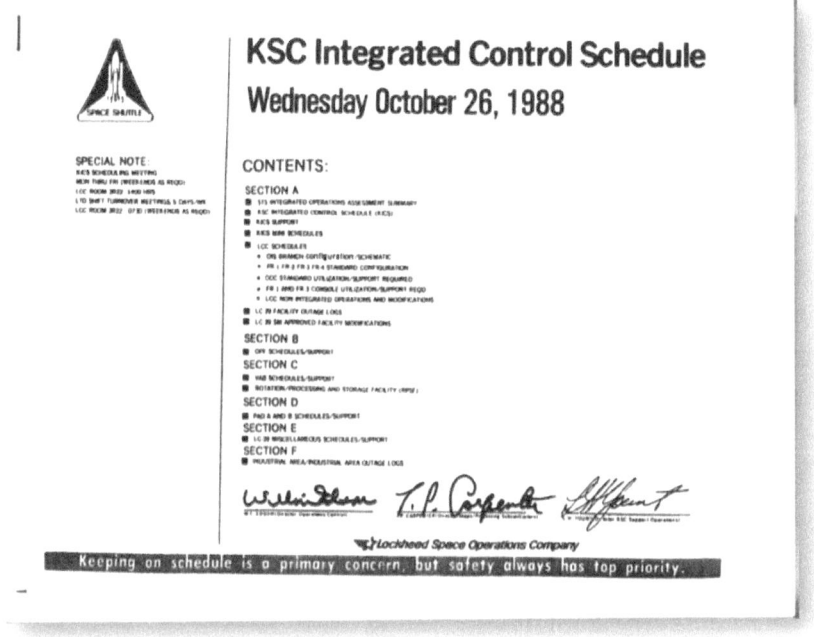

The schedule a few days after I first reported to work at NASA Kennedy Space Center. It was called the "KICS" for Kennedy Integrated Control Schedule.

These were schedules. Among the things I learned that first week, this place is big on schedules. Months later, I was told to cover the daily scheduling meeting over in the Launch Control Center. "Daily," as in there were two a day. Why have a single meeting when you can also have a pre-meeting. I looked forward to covering the meeting, which seemed to me, an easily impressed grunt at the time, an important meeting in an important room with important people. The meeting room was once wall-to-wall schedules. There were huge uninterrupted walls, front and sides just for this. The black and white lines and grids reached where they would have needed a ladder and a thousand magnetic labels back in the day. The schedules were not kept on the walls anymore here, though, everyone walking in with their trusty "KICS" hardcopy. Going to church? Bring your bible. It's part of the look, I learned.

After the first time I covered the meeting came many more. Some years in, everyone getting caught in a rainstorm, I rushed drenched to some random seat in the back. Usually, there were assigned seats. Today everyone ended up scattershot. In a very hierarchical arrangement, seats at the table were labeled. Seats around the table seemed to go from more to less important, with the nose bleed section three rows back. This place was big on schedules *and structure*. By chance, this time we were days away from a launch, and there was a major mechanical problem on the Shuttle. This is not to be confused with the usual brokes. This was a big one.

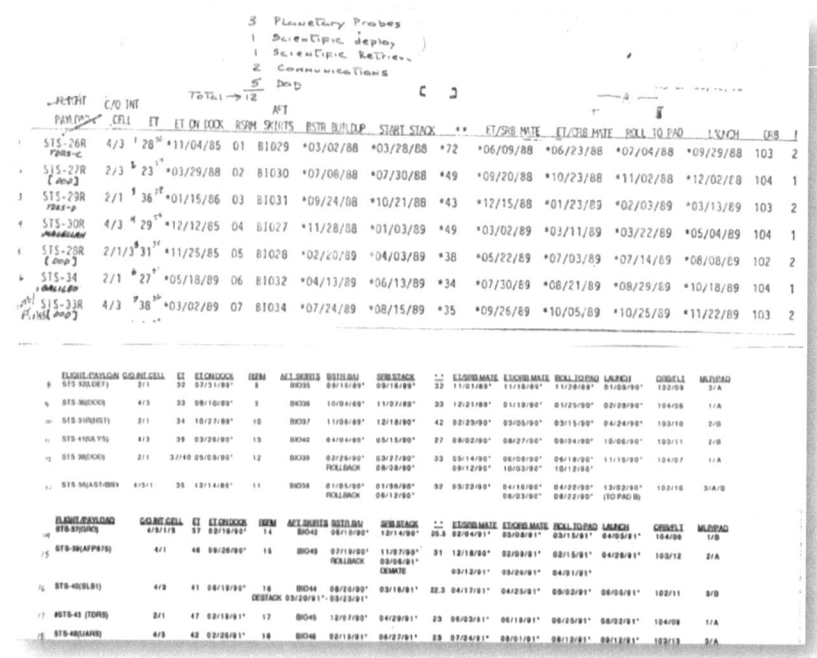

The major events sheet at the start of the daily schedule.

Next to me, by chance, ended up sitting our public affairs representative. It was pretty clear we would not launch on

Friday, and telling the world would save a lot of tourists a lot of trouble and traffic. I leaned over to the public affairs rep with a whisper about how "I guess it's time to tell the tourists it's no go." The reply – umm, not quite.

The team was told to troubleshoot the problem, come back later (another meeting, same day), and re-assess tomorrow (the next meeting). In the meantime, *we are still working toward a launch on Friday.* So, another day came and went. Then another. As said earlier, the repair was very likely and would take many days. The tourists never got word until late the day before the canceled launch.

Schedules were made to be changed. In government, given even more years, episodes like these would happen again, eventually, a supervisor telling me, "We say what we do, and that's all." To add in the meaning of what we do, how what we do adds up or not, was above our pay grade.

Above our pay grade? Tell me more. Intriguing, but incomplete. Of course, I went like Alice down that rabbit hole. Seeming a lifetime later, really just 10 years in, I had wandered into cost estimates in NASA programs. In 1999 I watched Bruce Willis as the psychologist on the big screen, who thinks he is helping a boy, only to find out – true, intriguing, but not the whole story. More so, I had joined a society of unreliable narrators. It was vast, pretty exclusive, and did a lot of math. The math part is why I fit right in, I suspect (I'm pretty decent at math). But of course, there was much more underneath the math.

"Optimism" is perhaps one of the most common words in NASA next to "cost estimates." (It appears 26 times in an Inspector General report on the subject). We didn't get paid to discover problems. We got paid to find solutions. Projects reported what they were doing. Report on plans? *The plan is to launch on Friday.* There is even a school of thought where estimating costs for NASA projects is not only thankless but worse, it's impossible (This is part of the "you can't nail jello to a wall" school, "endogenous" effects, and all that.) "It's a process" was a first cousin. Being too-close followed as another theory, but projects had a way of making sure independent assessment, from afar, would not happen. "It's a *team* process" – came the clarification.

As I write, I must remember all these experiences. The whale was huge, and malign, and NASA cost estimators are here to help that project, and we plan to launch on Friday. More so, it's worth remembering this in a world of fresh ideas, start-ups, technology and innovation. Since wild ideas are wild, until they are the baseline, perhaps we need to ask "who helps who?" among cost estimating and *new space* projects. A joke once upon a time in cost circles was "we need to build a reusable launch vehicle, so we can make better cost models." Cost estimating for space systems may want to help a project, like Bruce Willis wants to help the boy, but it should worry about its own state too.

THE SUPPLY CHAIN CRISIS: A HISTORICAL PERSPECTIVE

Need an excuse? Just say, "It's the supply chain." What first was an understandable shortage of personal protective equipment, because who could have predicted a pandemic, is now a shortage everywhere. The attempted, detailed explanations for the toilet paper scarcity at the start of the pandemic are now arm-waving followed by muttering "supply chain." Somehow, those two words are supposed to suffice. We once had Zeus, the fates, fairies, or divine intervention. Now, we have something even better to blame when things go amiss, the supply chain.

I can't help but recall the Space Shuttle and its supply chain crises. There were many, but two particular supply chain crises stand out. The first of these crises was underway by the late

1990s, and this hit close to home for me.

Our vendor was in town to check out a large valve assembly, part of the Rube Goldberg contraptions that connected the bright orange external tank to the Space Shuttles. I looked forward to the visit, having gotten to know the team well. During this visit, they gave me the bad news; their company was leaving the Shuttle business. After much cajoling, the best response I got was some vague assertion about how the job was no longer profitable. Of course, this boggled my mind. You make a one-of-kind unique part for the Space Shuttle, and it's not profitable? Huh?

After all, this was an era when making a part for the Space Shuttle meant bragging rights. This was so no matter what you made, even if just a bolt. You could do the mandatory advertisement in all the trade magazines, the Space Shuttle lifting off the pad, your company logo, with a very small picture of your product as almost an afterthought. Or no gizmo picture at all. If your product spun at 82,800 rpm, or jammed 75,000 horsepower into a pump that fits in a duffle bag, or some other technology magic (for the times), without coming apart, all the better. You had a right to proudly celebrate being a part of the world's most complex and capable flying machine ever.

Yet somehow, this was no longer profitable?

It's no secret that Space Shuttle contracts cover all costs. They covered even the cost of surprises. Which is to say your company never suffered from having failed to predict all the expenses that might arise. Then the government added a nice profit on top of all that as a percentage of costs. Even better, if you kept your government over-lords (like me) happy, there was an extra award on top of that – think of it as the federal government's version of tipping. How could this not be profitable?

It turned out this vendor was not alone in fleeing for the hills. Our parts people had been tracking vendors for years and were seeing the writing on the wall. By the late 1990s, many vendors had fled the Shuttle program like rats off a sinking ship.

Not being one to forget a question after a shrug and some simple answer, I investigated what was happening. How could not being profitable be the end of the question? Remember, a profit was guaranteed, and a nice tip too! A Ferengi could not be happier – the 300th rule of acquisition is "Always take any cost-plus contract, no questions asked."

A couple of better explanations came as I looked further into this – "return on capital" and, yes, the supply chain. It was the first time I would hear the blame being laid on this mystery called the supply chain. Eventually, I spent some years delving into supply chain management. That's a few years of my life I will never get back. When I left, it was for what I discovered. This tunnel had more dead ends than paths going forward.

What the figure – "return on capital" meant was just as much of a mystery. Accountants were having a field day here, but for the Shuttle program, the explanation could be boiled down to guaranteed profit. Yes, but it was not enough. Companies wanted *more* because shareholders expected more.

Translated into English, as a practical matter, a Shuttle vendor would tie up a part of their factory to make a widget for the Space Shuttle. Floor space was isolated, comings and goings controlled, and everything inside that fenced area was tracked obsessively. It could be the spore drive interface for the Starship Discovery or just a sensor. No matter the seeming importance, all the parts could be called anti-social, avoiding mingling with the rest of the people and parts in the factory.

Undeniably, that closed-off, very controlled part of the factory eventually made less profit than the rest of the plant. This was the late 1990s, so why build another plant or expand with interest rates at 7 percent? You could just kick out the NASA business and get the whole plant humming along at the much higher profitability of the rest of the plant. This was "return on capital" at work, if not for us launching Shuttles.

Our NASA Shuttle Logistics Depot would eventually tackle

this first supply chain crisis well through "in-sourcing," meaning getting into the manufacturing business ourselves. Also, *sparing* no expense. But it was not the last supply chain crisis in the Shuttle program.

Because it's always something, another supply chain crisis came along. As the space station program was completing hardware production, if not yet launched, even more vendors began fleeing. This time, it was for a new reason. So, if it was bad before, it would get worse. Vendors didn't like merely answering the phone on parts made years or *decades* earlier. The value to a vendor was in the making, not the supporting. The loss of Columbia the following year made matters worse – how to maintain a supply chain for whatever would come after a soon-to-be-retired Shuttle when it wasn't clear what was next. It's hard for a buyer to keep vendors around when unsure what you will be buying. Those thousands of vendors – those are links in your supply chain.

To make matters even more complicated, this went beyond the Shuttle. It turned out that small NASA programs and companies with no significant shift ahead were fine only as long as they could depend on big programs for parts. If a big program went away, well, think of those little fish that follow around a Sharks mouth. No shark? Big problem for little fish.

I eventually left the work on supply chain management. The field has lots of measures, two of which were efficiency and robustness. (The actual lingo was up-side or down-side flexibility.) It didn't take long to see the problem.

A very efficient supply chain cannot be very robust, or vice-versa. Yet the illusion modern supply chain management wanted to project was "you can have it all." Everything you might need, with the right software, standards, or model was just a click away.

In retrospect, the Shuttle program's supply chain skirmishes were part of a more significant battle across the private sector.

Imagine an engineer figuring out how to get the plant near

100% utilization. Anything less is wasteful. Modern supply chain management software keeps you near this hairy edge. When the surprise comes that no one could have predicted, say you need 10 more valves for the apocalypse, *you are not getting those valves*. Not any time soon anyway. You are too efficient to be robust. Everyone could respond quickly to a business downturn by shutting down a line or firing people. The reverse, though, responding rapidly to vastly more orders, was impossible.

We were drawing a trend-line nicely connecting the pile of mail-order catalogs to the internet to same-day delivery of groceries to the replicator on Discovery. All of a sudden, where is my instant gratification? While it would be too easy to channel Gordon Gekko on greed and all that, a more complex rhyme and reason may also find supply chain crises are increasingly of our own making.

❊ ❊ ❊

Today's start-ups in a world with near-zero interest rates won't be thinking about return on capital for some time. Supply chains are nonetheless being dismantled and re-assembled as we speak, courtesy of a pandemic. It's always easier to destroy than to create, to fire than to re-hire, and along the way to forget rather than remember. Perhaps your company's rocket engines will come from some other company, or you think this is a good time to just print it all in-house. As has happened before, large NASA programs will complete development and production blocks will follow.

Smaller companies will benefit from these supply chains around them in ways they may not recognize immediately. In all this, rethinking what we value in a supply chain, *especially people*, and how we balance it all is not a new idea.

While seeming an impossible ask, it is still the desire - to have it all. A robust supply chain has margin, valuing people, for a capacity to respond to unexpectedly increased demand. Yet this extra capacity is erroneously seen as waste. This was one takeaway from my foray into this field years ago. Robustness is not a defect, sub-optimization, or over-capacity, the same way reusability is not poor mass management; *it's a feature.* It's easy to say no one can plan for the next crisis, except they seem to keep coming rather predictably. In all this, perhaps the lesson is "robustness, for lack of a better word, is good."

PLANNING AND SPACE EXPLORATION

...and for development and commerce...

Back in 2007, the NASA plan was to go back to the Moon by 2020. This is not to confuse anyone with the current plan to return to the Moon by 2024, which might be 2028, or sometime later. Instead, this was the older plan as NASA launched its Shuttles on their last missions. Except there was a problem then, dooming that plan. The problem was still there years later, as the NASA administrator admitted - "we don't have what really would be a valid strategy that the common man would accept." This had hints of Albert Einstein's saying - "If you can't explain it simply, you don't understand it well enough."

NASA's lunar plans around 2007 were multiplying like rabbits. Among my jobs at the time were some far-term cost estimates, meaning far from the big show fighting fires in the center ring. Still, everyone playing my role of a cog in the machine was asked what they thought about the bigger picture. These were always valuable opportunities for anyone to offer their 2-cents.

- Return to the Moon no later than 2020

NASA, 2007

By then, I had learned a thing or two on providing feedback to projects, especially how no one ever got their suggestions adopted if there were too many, too detailed, or too far away. Obviously, we were told there can't possibly be that many errors, details would shake loose along the way, and anything far away will come with time. That's all a project's way of telling everyone too much feedback must be nitpicking, and the review is merely a formality. It was also a handy way to reject about 99% of the suggestions. Having seen this behavior before, I stepped back and pondered the whole plan. There was an error glaring right at me - where was our outpost?

If your goal is a sustained presence on the Moon, for various reasons, you should start work on an outpost just before the first landing in 2020. Technology takes time, and your plan should help keep your eyes on the prize. That avoids having all the goals down, but only half a plan. Amazingly, the outpost was missing in more than a few places. So, that was my comment, but prettied up in formal lingo and submitted to correct what was probably an oversight. More likely, the reply would be a name, enlisting me to help fill this in. That was also another project behavior bordering on cliche; beware of identifying problems, as you will be drafted to resolve them.

At one point, the comment was called "out of scope," a government agency's way of saying they only wanted comments on what was in the plan, not what was missing.

> The Administrator shall establish a program to develop a sustained human presence on the Moon, including a robust precursor program to

NASA, 2007

Regrouping, I could not help but wonder if what we were looking at was some mild plan. Or maybe these were goals, with plan-flavoring? Perhaps this was *plan-ish*. The word "notional" was spreading like black mold on NASA charts. Was this a notional plan?

Months later, work on a lunar outpost starting just before a first landing in 2020 was added to the plan. The single chart could easily be confused with a "big plan," showing the cost of everything up to the first lunar crew landing. This was just one of the Constellation program's many "sand charts," with the atrocious primary colors showing layer after layer of money and time. Some of these escaped into the wild. None of these plans persisted for long, but the notion that a long-term plan did not need to plan too far ahead would linger.

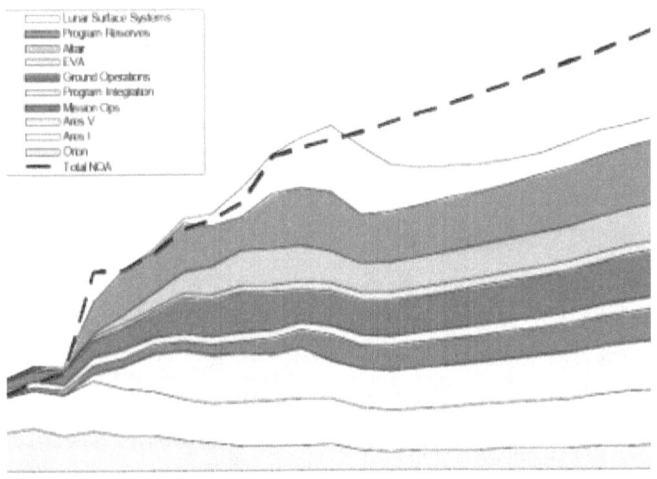

NASA, 2008

This puts some context behind the remark from our NASA administrator years later - "we don't have what really would be a valid strategy that the common man would accept." That "common man" would want to see the essential pieces, like a lander or an outpost. You might also ask where the spacesuits were if you knew enough to be dangerous.

In defense of the process, a "plan" is many things to many people. If you are of a certain generation, the jokes were about the Soviet Union's 5-year plans. Potatoes anyone? Or rather, fewer potatoes this year than last. This was more about people and what drives us to get things done. Today's bad press about plans comes from start-ups, espousing their virtue and an accompanying bias for action. Then comes the anecdote about SpaceX, in which Elon Musk reminds a new employee that they don't need to plan. They need to "just do things."

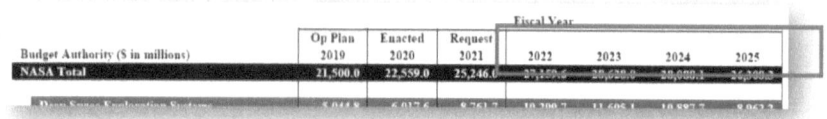

NASA, 2021

Ironically, a government agency can't help but plan. After all, budget documents look out for 5 years. If the budget does not arrive according to plan, the little-known secret is it's not that difficult to quickly re-plan. This mostly means delay, meaning pushing tasks to the right on a schedule, given the approved budget is concrete that is now setting, leaving little wiggle room. Unfortunately, this simple year-to-year adjustment of plans does not happen. The reasons are many, and there's no denying the complexity of what's involved. That would be like denying the complexity in some physical principle like least action. It would also be pretending we understand that principle enough to explain it simply.

This meanders into the confusion between a plan and planning. In my time at NASA, I grew fond of the term "situational awareness," a recognition of how little a project really controlled, all while being asked to put on the smiley face of optimistic goals. Situational awareness was my way of saying, at least there would be planning, which is everything (I imagine even at SpaceX.) This is not to be confused with plans, those transitory snap-shots of moments in time, two parts optimism and one part fact. Situational awareness is all about planning, knowing where you are, acutely aware of your surroundings, driven by a desire to "understand it well enough."

Eventually, I showed more patience than most for gathering up people and their numbers and adding them all up in a language that simulated English. So, I was asked to provide a formal "strategic analysis" of the lunar plans circulating in 2008. Remember, if your comment is about a problem, you get drafted to fix the problem. Projects use this as a deterrence to people giving too much advice. It's an effective tactic, except for a rare few who were raised to never let go of a question once asked.

My attempt at offering situational awareness was not received well. For an audience that would fit in at the Vulcan Science Academy, I found that while everyone agreed I had stayed in character, leadership had not (not even close). To top it off, sometime after, I would be the sole dissenter when a verbal roll call was taken on proceeding with one of the plans, the one about landers. Only in retrospect is this funny in the same way I laughed as a child at Wile E. Coyote with the train coming – that does not stop.

The Constellation program, with its lunar plans, was canceled a couple of years after my presentation. But the real casualty of this cancelation was planning.

This is an excellent example of learning the wrong lesson. Shoot the messengers. Long after this, asking our leader in spaceflight about this, his lengthy reply ended with how there was "no advantage to having a real plan." If there were to be a bonfire again one day, the thought went, the NASA program would not light the match, and even less, provide the fuel.

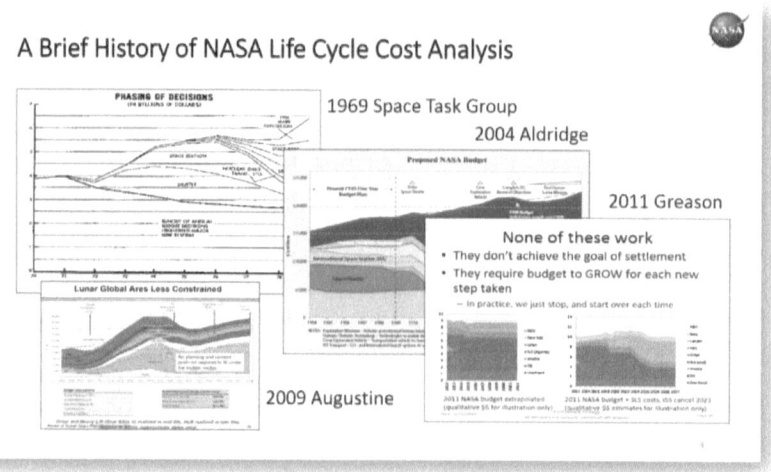

From NASA Human Spaceflight Scenarios, Do All Our Models Still Say 'No'?, my 2017 look at plans, planning, and scenarios for the future of NASA human spaceflight. In the scenarios where NASA models say "yes," prior steps must significantly drop in cost once operational to make the next steps possible. Otherwise, there are no next steps. And there are many steps to exploring the ends of our solar system. And one day, beyond that.

Soon enough, 2024, or 2028, or soon after, will imitate the oncoming train illusion. The dates will seem far away, then suddenly, they will be on top of us with our lunar plans, public or private. As with Mars, a generation away when I arrived at NASA, the decades fly by, but the train does not arrive. Admittedly, across teams, centers and agencies, NASA, and industry, I never heard it said plans or planning were necessary to know a goal was unlikely. On the other hand, good planning to get us closer to our goals can't happen if planning is seen as just bad news. Instead, real planning is possible if it's embraced for its advantages, increasing awareness, and fostering an unflinching look at our challenges. These challenges span space exploration, development, and commerce, all linked. Along the way, planning might even help understand what's happening well enough to one day explain it simply.

UNBURYING THE LEDE

It is 2022, yet as happens more often than not, NASA's budget is about the same as 2021. This is not new. It is a line likely to be reused next year and many years from now. It results from the usual lack of congressional consensus affecting most government agencies. Not that timely agreement would alter the basic math, where NASA's yearly budgets fail to make up for inflation, even when nominally an increase. And yet, even that shortcoming might be taken as a minor community note considering NASA's ambitions. Now, every step beyond Earth, in so far as NASA's contribution, can only come if every prior step is drastically cheaper and more routine.

This idea rests on the NASA budget for context. If lower costs are the picture, NASA's budget is the frame or the straitjacket, depending on your view. If NASA's budget only goes up a little every year, the cost of getting to Earth orbit must drop before going further. Otherwise, where did the funds for the next step come from? The idea repeats that a step beyond the Moon is possible only after living and working on the Moon is much more affordable than at first.

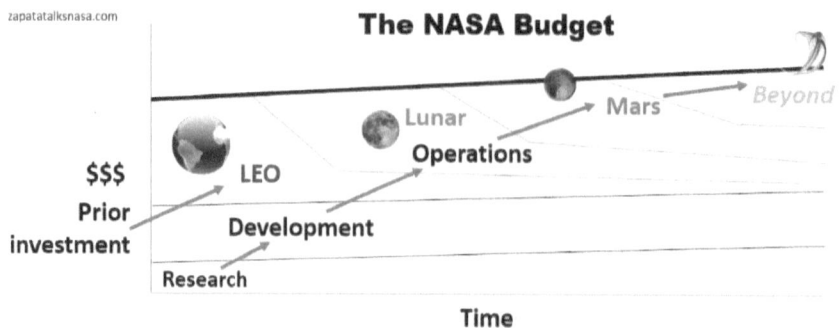

Each step as we explore beyond Earth can only happen if the previous step becomes much cheaper and more routine.

Living and working in space is a step in the middle, as we have done since November 2000 at the International Space Station. Each step, getting to and staying there, must get cheaper to free up resources to go further. Otherwise, plans to go into deep space are like asking directions and being told, "You can't get there from here."

All the while, NASA's budget has seemingly shot up like a rocket. But take the red pill, get your location, and it has actually gone down. Inflation provides the reality check. Yet even this view is generous. There are holes aplenty in assuming the cost of a *constant set* of NASA-ish aerospace goods, parts, services, research, and technology have gone up over time only as little as the official NASA inflation index says. Nonetheless, even using the (optimistic) official NASA inflation index, NASA's purchase power has declined by about a fifth since 1995.

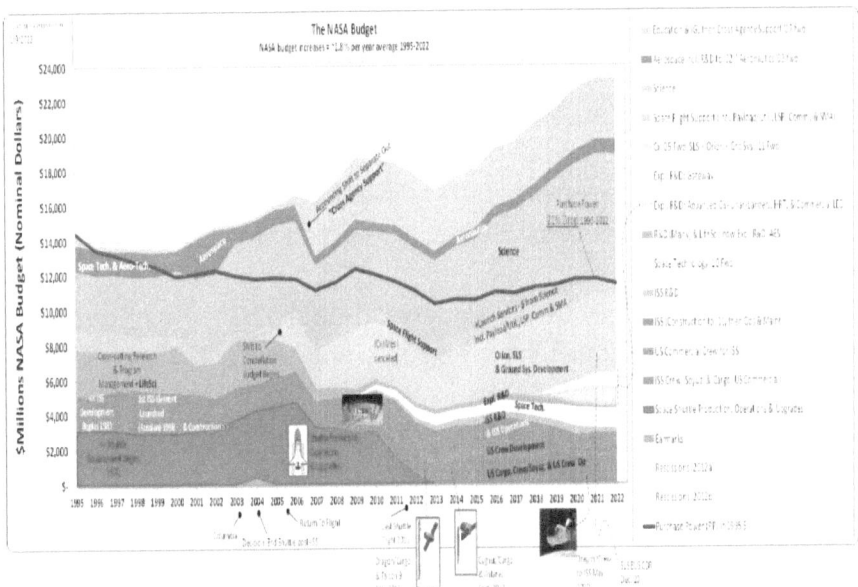

The NASA budget since 1995 has lost 21% of its purchasing power.

There are different paths to making the cost of each physical step drop dramatically, from getting to Earth's orbit to going further in relevant timeframes. (What feels like relevance being another matter.) But first, to be clear, these are costs dropping to NASA, a limited form of a much bigger question. Second, simply diverting budget resources away from one step to the next would look nearly identical in the budget graph. And therein lies the rub, as jumping ahead must be accompanied by another question – how best do we sustain our leap outward as we explore?

In one path, innovative public and private investments ensure costs drop in steps to the left (getting to orbit, staying there). At the extreme, costs (and so prices) drop so much that NASA can do even more in low Earth orbit for a fraction of what it once spent. In this model, NASA might have dozens of its astronaut researchers in rotation one day, getting there so cheap and staying aboard private space stations as more of a scheduling issue than a matter of budgets. Prior development, both public and private, can make this future a reality, conceivably soon. If a decade is soon anyway. The budget graph would look like the figure above, and it's easy to see the staying power behind each step. Here, the orbital economy is strong, making life better for people on Earth, from new medicines to materials. Once maintained by a thin thread back to Earth, the outposts are now towns with diverse, robust connections and supply lines.

Alternately, there is the outward dash. Here, the idea that every step must cost less along the way outward transforms into "we will spend less on each step." A military strategist will relate, considering how to grab the far-off target of value so quickly, with such surprise, that there is no need to secure the rear or strengthen your supply lines. You are moving too fast to stop and reinforce steps along the way. That is until you pause to ponder how to hold on to your gains. This strategy assumes speed succeeds, so the outward dash is not overcome by events – an inevitable outcome if moving too slowly. (Or so much for the dash part of the outward dash.)

For sure, NASA will get a new budget year after year – and the usual news will be about who's up or down, delays, "mitigation," and more inside baseball talk. Time will tell if the idea about each step getting cheaper and more routine –*as a requirement*– proves to be the reality behind the talk. For now, the optimists may see a budget half full, even if the container is getting smaller.

A REVIEW OF THE ASAP REVIEW OF NASA

Predictably, reports by committees read like a meeting with a few people speaking all at once. Why say something once, directly and simply when saying it five ways keeps every contributor happy their suggested sentence was not deleted? Nonetheless, this year's report of the NASA Aerospace Safety Advisory Panel (ASAP) is a refreshing read, saying many things rather simply, directly, and often. And these are things that need to be said. The lingo is the usual aerospace-speak, but the message is still rather clear. In plain terms, space exploration is changing as the world is changing around NASA, and so NASA must change as well. This change is necessary for NASA to continue to matter. The time is now. NASA's budgets are what they are, limited, and likely to remain so for the foreseeable future. And all this is especially important if NASA wants to go further than ever before, to the Moon or beyond. Increasing ambition and stagnation don't do well together.

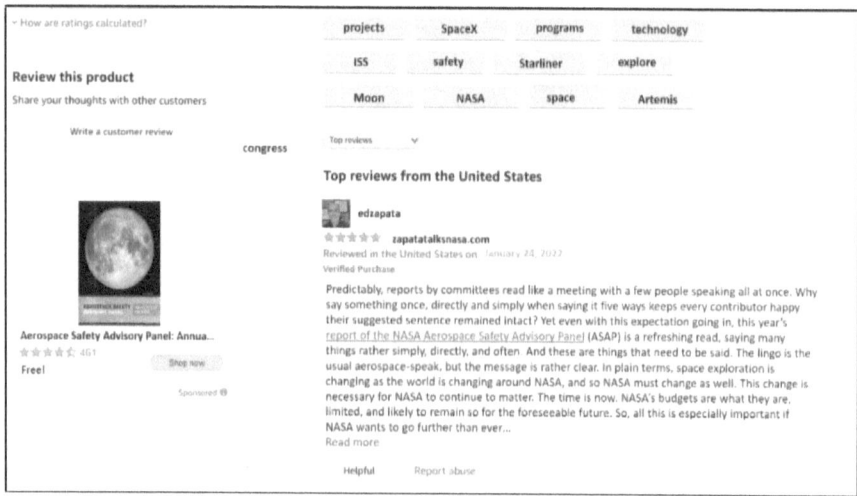

When it came to assembling reports by committee, I was usually too slow to step back from the line, so it seemed I had volunteered instead. There are always difficulties from the start in assembling reports on what can be dry, obtuse topics. It's only worse when cobbling together the sentences and never-ending edits while reaching for a coherent whole as everyone's focus is just a part. Having "assembled" a report was usually a fail. My goal was to have written one instead (which is no surprise to those reading this right now.) By chance, my last major task at NASA (before retiring) was compiling a report for DARPA on the XSP program's lessons learned. So, it's easy to appreciate the difficulty of keeping the kernels of wisdom simple, direct and on target – and I admire it when it happens, as with this ASAP report.

Ironically, a repeated criticism of the committee is NASA running it's major programs as if by committee – "the Artemis campaign is not established formally as an "Artemis program" … In other words, there is no clearly defined leader of the enterprise." Recursion is a thing, as it has fallen to a group of experts to note that at the end of the day a mission to the Moon should have leadership, not groups.

There is no lack of ideas about the changes NASA must

embrace. Queue Dylan - The Times They Are A-Changin'. The ASAP report avoids getting into these changes too much, instead focusing on a more fundamental question – *why change?* Previously, I've written about *what* might work, and *how*, if NASA's mission to explore space is to go further, faster, sustainably – and stay.

What is necessary for NASA to go further on its budgets (which inflation has eaten into over decades)? Each step we explore beyond Earth, first getting to orbit, then staying, must get cheaper to allow any next step.

How might we make this so? NASA went about getting cargo to the International Space Station after the Shuttle ended with public-private partnerships, to great success – resulting in the Falcon 9 and Antares launchers, and the Dragon and Cygnus spacecrafts. These were all had for bargain basement costs (as NASA costs go.) The plan is to repeat this approach as we think about keeping people in space when one day the International Space Station ends.

More so, the ASAP report repeats a mantra about a need for strategy at multiple levels, the importance of asking *"why"* NASA must do what it must do, including adapt, innovate and change. If a cart is being pulled in many directions, it's unlikely to make headway in any. As well, the ASAP points out this is unsafe. Or as the ASAP puts it – *"Ignoring the external forces and environment in which the Agency must function will place NASA in a tenuous position going forward, which in turn will impact how safely and successfully it will be able to carry out U.S. government missions in space."*

None of this is new. It took about 2.8 minutes after we would start to review a programs lessons learned document or other report providing advice to a program for someone to say *"you mean lessons not yet learned?"* These were always opportunities to be constructive, to ask "why" an issue arose, and to ask "why" it had proven untenable – and not for the first time. "Why" is always a good starting point, but it's just as valuable to ask at the end of the day as well.

TWO REPORTS, TWO NASAS?

"*What if we modified the main deflector to emit an inverse tachyon pulse, that might scan beyond the subspace barrier.*" In a pinch to explain something complex? Use technobabble, courtesy of Star Trek (TNG). Remember, there is no limit to the functionality of your main deflector. Fortunately for Trek, it's easy to follow what it all means as the crew pushes buttons to save the day. That or a wrench that looks suspiciously like a tuning fork also does the trick. Quickly translating odd phrases into action is one way to move a story right along, without making sense of what Picard just said.

In the real world, though, if space is hard, translating NASA-speak into English is even harder, by 3.14 orders of magnitude, with a confidence level of 95%. Two recent NASA reports just beg for a poor translation. However, if you know Klingon and want to shift to English, you know some concepts just don't translate. Still, there is value in the effort.

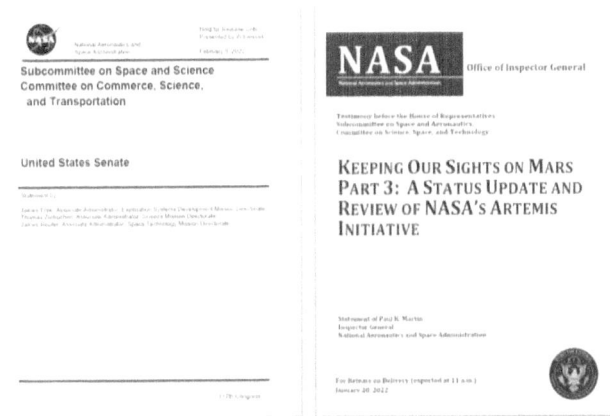

Two recent NASA reports, a study in contrasts.

Technobabble is not the sole domain of NASA. Even writing about technobabble falls into the trappings of technocratic speech. Take the statement *"the function of technocratic discourse in public policy is to advocate and promulgate a highly contentious political and economic agenda under the guise of scientific objectivity and political impartiality."* (The authors make an excellent case.) Still, we could use a dual-screen. On the other screen, a translation, "if you can't explain it simply, you probably don't understand it." A second translation below that, "beware people talking in circles."

This week, the first NASA report is from Inspector General Paul K. Martin summing up NASA's progress on its Moon program, Artemis. *"Moreover, our detailed examination of Artemis program contracts found its costs unsustainable."* The curious part of the report is that one word – *unsustainable*. Such harsh language is usually scant in reports by someone inside NASA, but outside the NASA project they are critiquing. Still, some possible translations for "unsustainable" help simplify. Not that there is not a dictionary definition, even one that offers a good analogy, like *"upsetting the ecological balance by depleting natural resources."*

But some more straightforward translation is still possible.

Unsustainable as meaning *"this won't end well"* is just four words, none more than one syllable, and capturing the feeling. A literal Spanish version of this, *"eso no acaba bien"* is more substantial, with heart, and closer to what unsustainable is all about here. Another translation comes courtesy of everyone's favorite engineer, so much more human-seeming than us real ones, where unsustainable comes out the universal translator as Scotty saying, *"I cannae change the laws of physics."*

The inspector general also provided numbers – *"our estimate of a $4.1 billion **per-launch** cost of the SLS/Orion system for at least the first four Artemis missions."* Translating this one is tricky without seeing the math. Inevitably there comes a year when a NASA project says it is operational, in a broad sense, unrelated to detailed dollar flows and mind-numbing accounting. A simple translation, first going to Klingon and then to English, to best capture the feeling on hearing "four-point one billion per launch" is *"take all the budgets for all this stuff, add them up, and divide by the number of launches in that time, and you get about 4 billion."* Some translations just fail. Setting the universal translator to "simplify" spits out a better version, just three words, again one syllable each – *"that's not good."* The IG goes on to talk about cost-plus contracts as the cause of such high costs or, for the uninitiated – assorted ways of saying *"never buy in a sellers' market."*

The second recent NASA report provides a more positive note. It was published with little fanfare compared to the prior IG report. Here, NASA Associate Administrator James Free provided testimony to the US Senate subcommittee on space and science. He focuses on the *"New Space Environment,"* but this has nothing to do with space debris and everything to do with partnerships. *Partnerships* can be that deflector shield of late, handy in a pinch but leaving questions about how that works. There must be something about exploring space beyond the challenges of expressing it, as the excitement, inspiration, and awe are left behind when trying to answer the question – how are we doing?

Perhaps space exploration is also ineffable here on the ground, but saying that would be too easy. Partnerships are essential in the search for how unlimited horizons meet limited budgets. So, what are partnerships, roughly translated? It would not be oversimplifying to say a partnership is NASA (public) and a company (private) "*moving forward together*." Public-private partnerships align everyone's incentives in ways a "cost-plus" contract cannot. This makes sense. Real partners sink or swim together, and how could we envision exploring space otherwise?

The contrast could not be starker when comparing the testimony by Associate Administrator Free and that of NASA Inspector General Martin. We call engineering about a month apart to ask, "How are we doing," only to get two very different replies. One, a sense things are amiss, another saying they are on it, and they have an idea. The unknown possibilities are around the corner, and it's time to see if we can modify that deflector shield, queue action, and move forward together.

NASA - JOIN THE CLUB?

Groucho Marx famously said, "I refuse to join any club that would have me as a member." Here, the search for where you belong is not just a horizon you can never reach but one you don't want to. As the world changes around NASA, there is no lack of similar questioning – what is NASA about, where is it going, and if it got there, would it want to join the club? The purpose of space exploration can be a bingo-card debate marking off everything from why to how with who's in the club thrown in for good measure. Blurring roles and objectives, mission and mandate, rather than recognizing these as distinct, creates more confusion.

> *Purpose, noun*
> *-something set up as an object or end to be attained, resolution, determination*
> *-a subject under discussion or an action in course of execution*

Atop the blur is a budget reality. Last week, NASA's budget for 2022 was belatedly approved, giving NASA 3.3% more funding than last year. This is in a year where inflation may well run at least twice that. It's not just which club you want to join, it might be you can't afford to join when you get approved.

Oddly, purpose does join the club, at least as the dictionary would define the word. An "end to be attained" has an ending. This view is at odds with a sense of purpose as a process that never ends. This is befitting, in a universe with no end. The exploration of space must be a journey, not a destination. Until we go down to how projects as action measure up against real purpose.

By late 2009 it was clear that NASA's Moon program, then called Constellation, was not long for this world. The enthusiasm of just a few years earlier was gone for many, myself included. For others, there remained a strong sense of mission. I was an early advocate, as far as getting it over with. A simple throw-back using Space Shuttle hardware would prove a practical maneuver, a quick and dirty pit-stop given the Shuttles pending retirement. But in just a few years, budget realities and vagueness of purpose showed the fatal flaws of such thinking. Still, admitting error and exercising your right to learn was not in the cards for everyone.

Over in Huntsville, Alabama, a Marshall Spaceflight Center engineer expressed a typical frustration – "if NASA is not building rockets, there is no point to having NASA." Following the breadcrumbs in a debate about NASA's purpose in life finds fundamental disagreements over the meaning of purpose and mission.

If you believe building a rocket is your mission, you end up with a very different worldview from someone who thinks the task was to get to orbit over time more often and ever more easily. In turn, the later view gets usurped by science, where the real purpose is learning and the "aha" moment. Here, any rocket and everything about getting anywhere is just a means to an end. The only real purpose to it all is scientific discovery about Earth, our solar system, and beyond.

Administrators are not immune to getting pulled into this debate about NASA's mission and purpose. Ask a NASA administrator about NASA's mission and purpose. Don't be surprised if it's the moment where the candidate is asked why they want to be president. In 2010, NASA Administrator Bolden, asked about the shift to commercial launch, said, "my vehicles today are commercial vehicles." In 2021 former NASA

Administrator Griffin bemoaned that if NASA "seems to be evolving into a government bureau whose job it is to write checks to billionaire entrepreneurs," then "there really isn't a purpose for NASA."

Yet, in both views, most NASA dollars have always gone and continue to go to private companies. In the one view, a changing relationship between NASA and the companies it contracts is merely a necessary twist, but the shift is an unwelcome storm in another. In one view, if the mission was never about control, the mission has not changed much at all. In another, control was the point. As a NASA Ames Research Center engineer was fond of saying when this topic inevitably arose - "you give up control, for something better." No one fighting a war would ever say building ships was the goal, except in a view at the dock. A means and an end can be close, but it's one that obviously leads to the other.

Accountability and responsibility rear their heads here, confusing matters by seeming to be one and the same. Yet anyone who has ever hired a contractor knows they may hold a company to account for a job well done, but it is you who are responsible in the end. Your home is yours alone and your responsibility, even as you may call others to account to get a job done.

There is no lack of official views on NASA's mission and purpose. Learning goes with any scientific and research organization, for NASA more now than when it began. This harkens back to purpose as a process. Any end is always a moving target. Yet if your goal is to lead, the matter remains - how do I know where the front is? The pointy tip of the spear is a rocket to some, a landing to others, or none of these if like a Trekkian V'ger you must "learn all that is learnable."

Any view on NASA's purpose and mission will resolve where the devils in the details. An insider's view is that of an ongoing battle, between the new world of commercial partnerships, firm fixed price contracts, and a focus on buying results versus a previous mindset of cost-plus contracts, paying by the hour for maybe something. Partnerships at the least accept a budget that is what it is, trying to figure out how to get ambitious results regardless. As far as being productive, that's better - if risky – than insisting (or whining) about the good ol' days and how such budget pressure shows a lack of commitment by Congress. Choosing to do what merely may fail to achieve your purpose beats doing what is guaranteed to fail.

Though among the details, directed work is where a sense of NASA's broader purpose and mission can run aground. If you are given orders by Congress, these orders become your purpose. Not having worked backward from an overall mission statement, the chances the directions take you to any goal are dim. More so, the orders include how, where, and by who, as with many a cost-plus project (the SLS and Orion systems). Here the means becomes the mission, and the purpose becomes an output.

The space economy is taking a further hit, events with the Russian invasion of Ukraine adding problems and uncertainties. One NASA purpose has been to bring peoples together across the borders we do not see from space. For one relationship here, that mission is hardly ripe to flourish. As NASA budgets have effectively dropped in 2022 (after inflation), adding to the changing environment, it's an excellent time to ponder NASA's mission and purpose.

Understanding our planet and exploring our solar system and beyond has a point. Improved aviation has a point. Space stations where we might one day produce unimagined goods, from medical to materials, have a point. Learning improves lives. The matter is NASA having the flexibility to reach for the goal, even if it's always trying to reach a horizon. And knowing and (mostly) agreeing on your mission and purpose helps along the journey.

If I move on, who does this?

> NATASHA ROMANOFF, AKA BLACK WIDOW, AVENGERS ENDGAME

Risk. Risk is our business. That's what the starship is all about. That's why we're aboard her.

> CAPTAIN JAMES T. KIRK, STAR TREK, RETURN TO TOMORROW

*The anomaly, my ship, my crew; I suppose you're worried about your fish, too. ... For that one fraction of a second, you were open to options you had never considered. *That* is the exploration that awaits you. Not mapping stars and studying nebulae, but charting the unknown possibilities of existence.*

> Q, STAR TREK: THE NEXT GENERATION, ALL GOOD THINGS

MONTY HALL, GOATS, THE ODDS, AND NEW ROCKETS

And what does any of this have to do with space exploration?

2004, a large, musty conference room at Kennedy Space Center, today holding only the six or so of us to mull over a question trickled down from on high. What is the *probable* year the Shuttle will complete another 22 launches? I'd been invited, having taken an interest in the topic of these probabilities and occasionally contributing some value. From my corner, the question is pretty cut and dry. The calculations leave little room for debate. But that's not to say today there wouldn't be a lively discussion. The launches in question were those to come once we resumed operations after the loss of Columbia. The mission called for completing the construction of the International Space Station, with the Shuttle program told to call it a day when done.

Launching at a certain pace is only one of many probabilities calculated by NASA's brain trusts. Much more analyzed, parsed, and debated was the probability of rockets failing. Even when failure is not an option, we know the possibility hangs in the air. Soon we will see this in play, the dice tossed for the NASA SLS, the SpaceX Starship, the ULA

Vulcan, and the Blue Origin New Glenn rockets, with more to follow. So, what do probabilities tell us to expect in the first launches of these rockets – more or less, and with what certainty? More importantly, what do the numbers mean?

> "THE ODDS AGAINST YOU AND I BOTH BEING KILLED ARE 2,228.7 TO 1."
>
> -SPOCK IN STAR TREK, DEVIL IN THE DARK.

Star Trek's Spock dead-panned the probabilities for any plan; the numbers were never good. Kirk either found ways around the odds or ignored them. Everything I needed to know about probabilities and what to do with them I did *not* learn from Star Trek. Nonetheless, I probably picked up a fascination with forecasting the odds from Spock. As our meeting about the Shuttle droned on, it too devolved into a tense Spock/Kirk dialog. I could tune out the unkempt room and poor lighting in a Walter Mitty moment, imagining the participants in Starfleet uniforms. Spock – "the numbers are clear." The Shuttle would likely do what it always did, launching about 5 times a year. Without flinching, pull that band-aid. Here is your answer. Always added was "give or take," all the jazz about if this, then that. Kirk responded impatiently. We must encourage the crew, err, workforce, to do better. We must "drive" the schedule, but the word "push" was implied. How to do this went unsaid.

Before entering the room, I figured the hour would be a curious break from my other duties in my day. *Fascinating.* Numbers can be stimulating, and I knew most people in the room for many years. Here came a chance to catch up and refresh on just where the numbers stood of late. Take it in. "Let the numbers wash over my mind," I thought. Sit back. My pre-disposition said listen to Spock and take his numbers as gospel. Even his guesses were better than most people's analyses. Not everyone jumped on board. From an expected Zen, the meeting went and zagged instead.

There is a phenomenon – Gambler's Fallacy – whereby a person thinks a prior event affects the probability of a future outcome. If the little ball on the roulette wheel just landed on 30 Red, there is a flawed belief that 30 Red is a tad less likely to land the ball again on the next spin. Naturally, we believe the past is stored in some ledger, ready to affect the future. Watch out, Kevin Durant has a hot hand! As our brains detect patterns helpful to our ancestors to survive the day, we store what happened and try to connect it to what will. We are our ancestors' children with practical pattern-matching skills, or we would not be here. It's all good for staying alive another day. But give someone a hammer and, well, don't look for patterns at the roulette wheel. Working the Shuttle's probabilities, we may not have focused on people's biases or misconceptions, but we did know how the wheel worked.

Besides Star Trek, where Spock says, "The odds against you and I both being killed are 2,228.7 to 1," a child growing up on early 1970s re-runs might catch Let's Make a Deal. You could do worse. Audience members chose between a sure thing like $500, ending the game and everyone's fun, or possibly something better. The car! Go for the car! Famously, a contestant having chosen Door #3 saw the host, Monty Hall, unveil Door #2. Then we saw a stack of tuna cans or a goat chewing hay. (As a child, I wondered if you actually got to take home the goat, not a bad prize, because who wouldn't want their own goat!) Then came the moment orchestrated to shift everyone forward in their seat. Monty Hall, the consummate gameshow host, easily kept the fun going. He asks the audience member who initially chose Door #3, now seeing a goat behind (initially unchosen) Door #2 – "would you like to switch to the unopened Door #1?"

Counterintuitively, switching doors is always the best move. The past event, like the ball landing on 30 Red, here showing a goat behind an unchosen door, has no effect on the future event. And the future event is a 2/3rds probability of a car behind either unchosen door. In this case, *that's the remaining door you did not choose initially.* Your initial door still has only a 1/3rd chance of being the car. Switch doors! Switch doors!

Because probabilities are not patterns, but our brains insist on connecting the dots, events playing out tax our brain's operating system. Monty unveils a door. The ball lands on Red 30. We think, erroneously, the next event must now have a different probability– because what are the chances of Red 30 twice? If a plane crashes into a home for sale, buy it. It's been pre-disastered!

> *"Honey, let's take the house. It's been pre-disastered!"*
>
> -GARP IN THE MOVIE THE WORLD ACCORDING TO GARP

More likely than not, NASA will soon launch its new Space Launch System (SLS). Around the same time, it looks like SpaceX will launch its new Starship in its complete configuration – unlike earlier tests with just the upper half, minus the booster. Blue Origin will come along with its large, new New Glenn rocket, and United Launch Alliance will have ready its Vulcan rocket (Spock would think the name a logical choice.) It's easy to imagine the rooms of people talking about probabilities of success vs. failure on that first launch. There are many asides about partial success, the first vs. the second stage, the launch vs. landing, the software or the engines, and so on. Yet eventually, we will know what is behind each door, one rocket at a time.

Who gets the goat? FAA guidance says a new rocket from

an experienced organization will succeed more likely than not on a first launch. A new rocket from a new organization will more than likely fail. For all four new rockets, all together, experienced teams and new teams, the chance at least someone has a bad day becomes a pretty good bet.

All this comes with asterisks about as long as the one's on your cell plan bill. This is *only* true over many samples, much more than just 4 rockets. Also, we don't live in all the universes where the samples can be run again and again. In my time running simulations at NASA, I was fond of running out 30 years, and running those 30 years 100 times, then averaging it all. In the real world, we don't get to see the other 29 universes like the Avenger's Dr. Strange. Car, goat, goat, car, goat, goat, goat, car and so on. Worse, we may not even see our own universe entirely play out. A two-handed economist has nothing on the likes of aerospace folk generating possibilities. NASA, SpaceX, and ULA have new rockets, except in the FAA asterisks, "A formal definition of "new launch vehicles" does not exist." The same might be said of new operators and experienced operators, legacy know-how, and when it's present or not.

This can all sound less than useless, more likely than not. The Shuttle program put astronauts on its first Shuttle launch. We knew long afterward the Shuttle program woefully overestimated its chances of launch success. Still, the optimistic probability supported going ahead with that first launch with a crew. Inversely, the Apollo program's chance of success for a first Moon landing was 5%. The under-estimated number was kept hidden from the public, and the program proceeded with the Moon shot anyway. Spock spits out the numbers, but Kirk just goes ahead with the plan.

With time, there is an awareness the use of probabilities lies in recognizing the limits of small numbers. SpaceX has recently been launching once a week. Who knows, a launch today, a launch tomorrow, before you know it, you might have regular launches once a day. We must know where to change the game, between spins of the wheel. Knowing your probabilities tells you where to invest, but also the goal – *to get the odds forever in your favor*. Learning loves big numbers, not tens and hundreds of spins of the wheel, but thousands, and millions. Admittedly, NASA is not in the business of big numbers, but the recognition is there that if one day we have space travel as vast as air travel, NASA's research and development investments will have played an important role. Of this, us number crunchers are quite certain.

NASA, AEROSPACE, AND OPTIMISM

In search of the right setting

It's not surprising to see studies again showing optimism can help us live longer. There is a circularity here. Any news about being optimistic and living longer promising to live on quite a while. Good memes, by definition, persist, going from trending to chitchat, back to studies, and then appearing again in the news. There is no shortage of reasons why optimism and living longer go together, at least for individuals. Lower stress or adopting healthier behavior are among a laundry list of explanations. Yet what of organizations and optimism, like NASA?

> considerations to achieving operational success. In fact, many people we interviewed raised the "Hubble Psychology" – an expectation among Agency personnel that projects that fail to meet initial cost and schedule goals will receive additional funding and subsequent scientific and technological success will overshadow budgetary and schedule problems.

> [Lessons from NASA Major Acquisition Projects]

> NASA's major projects will a[lways face] integration risks because they [push] the state of the art in space te[chnology;] management and oversight p[ractices,] optimistic cost estimating, bu[t also] contractor performance—are [...]

> qualified staff dedicated to cost estimating. Recently, for example, NASA's Inspector General found that "a culture of optimism and a can-do spirit permeate all levels of NASA, from senior management to front-line engineers. Although this optimistic organizational culture is essential for realizing groundbreaking scientific achievement, it can also lead to

> difficulty and complexity, technical uncertainty, stakeholder conflicts, scope changes, unforeseen events, and other not really unpredictable bad luck. Project planning is usually over-optimistic, so the likelihood and impact of bad luck is systematically underestimated. Project plans reflect optimism and hope for success in a supposedly unique new effort rather than rational expectations based on historical data. Past project problems are claimed to be

NASA optimism.

Now 64 years old and going strong, NASA has a special relationship with optimism. On the one hand, NASA is often accused (as if pending trial) of being overly optimistic. It's no wonder NASA projects cost many times and take many years longer than initially promised. Here, excessive optimism is cast as the culprit, alongside the scolding.

There is a flip side to optimism as a flaw. Here we embrace NASA's bright-eyed optimism to walk away inspired. "Failure is not an option" is the one NASA motto in popular culture. Or at least it is, thanks to some artistic license and Hollywood. Still, imagine if that Apollo moment had gone sideways. Your spaceship has blown a gasket on the way to the Moon. One more like that and she'll be blown to bits. There is the real risk of running out of oxygen or freezing to death, with no turning back, orbital mechanics, and all that, meaning this is not some movie where you turn around on a dime. This is physics, not fiction. The leader back on Earth says - "Let's all do what we can to return our people safe and sound. We know the outcome is uncertain, but blah blah." This version would not have won the motivational speech of the year award. A crisis is probably not the best time for blunt force realism. Instead, there was a much better pep talk, and two parts hard work and one part denial equaled success. Optimism lived another day.

NASA is not the only example of pathological optimism, or on the other hand, a necessary optimism worthy of praise. Wholly new start-ups, as well as new companies filled with experienced workforces, are also not immune to catching these variants of the Optimism-19 bug. The usual advertising is a high launch rate, a low cost, and a large, growing market. We will build it, and they will come – perhaps it's an orbital space tourism, factory, or something or other spinning in the sky. It has everything except investors or details and those pesky equations adding up. Yet when it does add up, we rediscover Arthur C. Clarke's 3 stages of an idea somewhere at the step about how "I said it was a good idea all along." As a NASA mentor once said, it's a special kind of engineer that walks into the room filled with manure on Christmas day and yells, "Yes. There must be a pony in here somewhere!"

The three stages of a revolutionary idea, according to Arthur C. Clarke:

(1) It's completely impossible,

(2) It's possible, but it's not worth doing,

(3) I said it was a good idea all along.

Yet knowing optimism can be good for you, or when it's terrible, is much clearer than understanding why. Here the reasoning gets murky. An evolutionary explanation posits that we ask what happens given the opposite? For individuals, this works. Optimism, even when it's a delusion, must help more than hinder. That is, pessimism is not just biting, it bites you. Except quickly muddying things, this leads us to the vague idea of many different lids for every pot. Otherwise, we would all be optimists, and the realist, and definitely the pessimist, would have long gone extinct. Asking why optimism persists in organizations only gets more complicated.

NASA and the space sector have no shortage of wildly inspiring, innovative and, yes, very optimistic ideas. We might put astronauts into a chemical sleep for a trip to Mars, reducing the required air, food, water, and living space. We might use fungus, fed with nutrients, to turn inert, rocky dirt from asteroids into rich soil for planting. Taking some fungus and nutrients beats lugging massive amounts of fertile soil all the way from Earth. Energy sources developed for space can add to a robust, clean energy portfolio back on Earth.

There will never be a lack of Theranos's going too far (will we end up at criminal optimism?) But 50 years from now, it's a good bet someone remains inspired by Dr. McCoy's salt-shaker medical thingy, and they are working to make it a reality. Like the Thermians in Galaxy Quest, the ultimate metaphor for inspired scientists and engineers, optimism encourages us to make things real, *even when they begin as a fiction*. Not knowing that putting people on the Moon in a decade was impossible, NASA did it. Not knowing that new launchers and new spacecraft for a dime on the dollar were impossible, NASA and its partners went ahead and did it too.

All these NASA projects, and so its ambitions, live inside a budget that, contrary to popular belief, has been decreasing yearly for a very long time. This is the reality of a budget that does not keep up with inflation. Since 1995 NASA's purchasing power has declined by about 20%. If NASA got $5 a year in 1995, today it is (really) getting $4. Practically speaking, you get a raise every year, but the rent goes up faster. In such an environment, the habitat for ideas, is it natural some projects mutate and become even more virulently optimistic than usual? Inversely, is NASA's new partnerships approach to most all new contracts in spaceflight an adaptation? Here we have the optimism that competition and ingenuity will save the day – and fit in those constrained budgets when nothing added up before. As the Red Queen said in Alice in Wonderland, "Now, here, you see, it takes all the running you can do, to keep in the same place. If you want to get somewhere else, you must run at least twice as fast as that!"

Having seen enough numbers in aerospace projects never add up, what's surprising is the widespread insistence to keep trying. (I was often the one who did the math.) Some of this drive is likely maladaptive, like turtle hatchlings – we do what we have always done. It succeeded before, so it will likely succeed again. But the world changes and that light the hatchlings head toward is now a beach house (and death.) Another part of this drive is very different, something hatchlings can't do. It's optimism as two parts hope and one part learning by trying. By learning, we can override some programming - correctly recognizing the Moonlight and heading out to sea.

Still, in 2022 no one can tell if NASA and the aerospace sector will find the perfect setting for our *optimism dials* anytime soon. (Bulls#*!--Impossible--Denial--Maybe--Difficult--Challenging--Achievable) But what is likely, is everyone keeps trying all the settings. And who knows, instead of doomscrolling, maybe we'll also live longer and prosper the more we fiddle with this dial.

MIND THE GAP

"We have one data point. All we need is one more and we can draw a line." This was one of our many meetings where we dwelled on lessons learned, the Space Shuttle, and what's next in reusable launch. As far as jokes go, at least for number crunchers, this was a good one. Except that as much as NASA invested in technology and concepts to follow the Shuttle, and we worked toward that second data point, no one could have guessed how it turned out.

From a viewing spot close to events, till about 2001, no one would have predicted the next flag-ship NASA rocket would toss aside the reusable Space Shuttle orbiters, keep the expendable solid rocket boosters and orange tank, and top it off with an old-style space capsule for crew. Although NASA always held around the idea of an extra-large launcher based on the Shuttle's expendable parts, especially for Mars missions, these usually lost out when analysts (like myself) provided advice on what was best for the future. In retrospect, justifications for investments singled out what was not ready, technically or economically, while promising to be much more so in the future. Inversely, not making the cut led to taking it for granted, erroneously, that a Shuttle-derived expendable presented no technical difficulties, losing out merely for predictably high operating cost. The devil we know was easy to discard. Though any sense of surprise was tempered by knowing that come decision day, the allure of the known would be back at the table and own the table too.

The real surprise no one foresaw is how even as the tide clearly shifted by 2004 and NASA was going back to expendable launch, by 2015, it ended up with an advanced, break-through reusable launcher - *anyway*. This improbable scenario came about courtesy of NASA's partner SpaceX evolving an initially expendable first stage (and NASA investment) into a vertically landing reusable booster. Just last week, SpaceX successfully landed and returned a booster for the 129th time (out of 140) after placing a Dragon capsule in orbit to robotically deliver supplies to the International Space Station. If an automated drone delivers stuff to your home one day, remember NASA did it first.

On top of these improbable shifts, nearly no one saw what was coming after the Shuttle's retirement - a ten-year gap in US human spaceflight. Though as a post-Shuttle gap became inevitable, it was pointed out that before the Shuttle's end, we had the end of the Apollo program and its gap. No US crew went to space between the Apollo Soyuz Test Project flight in 1975 and the liftoff of Columbia on its maiden flight in 1981, six years later. Eventually, the gap in US human spaceflight got its second data point, going *UP* from *six years* after Apollo to *ten years* after the Shuttle, even as it might have seemed that surely a six-year gap after Apollo was a fluke. We can't use that one post-Apollo gap data point to tell us anything about a possible post-Shuttle gap. Right?

So, NASA went expendable after the Space Shuttle, after so much time and effort invested in reusable launch and technology, but NASA got a semi-reusable launch system anyway through its partner, SpaceX, *and* we had no US crew launch capability for ten years. This was not quite predictable, and we had no lack of predicting.

As NASA makes plans for what comes after the International Space Station, the question comes up again - will the continued US human *presence* in space since late 2000 see a "gap" one day as happened with our ability to get there? Again, there is not a lot of data, which could be a blessing in disguise. We can put that next data point where we want, it would seem, as it's not that we have lots of data spelling out the inevitable. Or do we discard the meager data points at our own risk?

Since NASA doesn't end major space programs often, we could simply take away to expect the unexpected. Nonetheless, seeing why previous gaps occurred might tell us something useful about what else to expect. If we feel Deja-vu it might be for a reason.

NASA's plan is for commercial space stations with global customers and markets after the International Space Station comes to an end about a decade from now. NASA "intends to implement an orderly transition from current International Space Station (ISS) operations to these new CLDs (commercial low Earth orbit destinations.)" NASA ramped up funding for its partners here in 2021, and previous experience shows it's a wonder what NASA and its partners might accomplish with as little as $100 million a year when going commercial.

Yet relatively small amounts of yearly funding, as far as major NASA projects go, should also set off a yellow alert. The gap in US human spaceflight after the Shuttle might be blamed on a lack of funding to carry out or to accelerate a plan, but that is simply restating the problem another way instead of questioning further. Why is there a lack of funding? Better yet, it's worth asking if this is how everyone would see it – especially the people providing (or not) the funding.

If you stare at the NASA budget over time long enough, a first oddity is the lack of much *new* funding after the decision to retire the Space Shuttle. It's easy to say NASA received plenty of funding before that to get ready for this day. Aren't we done yet? Instead, it's the Shuttle budget that goes up after the loss of Columbia, naturally, to improve systems before resuming its last launches to finish the construction of the ISS.

Where then did NASA get money to do what would come after the Shuttle? Much of the money for what was to come next *came from existing funds*. In particular, knowing your current system is being retired, NASA in 2005 (relatively) quickly diverted its existing R&D capability away from Shuttle upgrades and other research towards the specifics of developing the next thing – a large expendable launcher and spacecraft. (It's worth remembering that before the loss of Columbia, but lacking the authorization to replace the Shuttle, NASA had a plan called "Shuttle 2020" – to keep operating the Shuttles clear through 2020.) But the efforts for what comes after the Shuttle really pick up steam only after the Shuttle ends – the next system being Shuttle derived after all, and using a now truly freed up Shuttle organization. Another good part of the freed-up funding post-Shuttle goes toward re-establishing operational capabilities to get to the ISS, today's commercial cargo and crew services.

What might these lonely data points in the history of major NASA programs ending mean for avoiding a gap in our human presence in space post-ISS? We are perhaps a decade away from that day, but a world post-Shuttle also seemed so far away for the longest time until it was right on us. Mostly, it seems any major NASA spaceflight program has limited runway for what comes next. Next steps require a combination of investing very little funding to set the stage and waiting for much more funding once the prior program ends.

There are some wildcards – we see the new line items appearing for the Gateway, a lunar lander, and for what comes after the ISS. It would seem the post-Shuttle world continues to lead to a more diverse NASA portfolio with smaller spaceflight programs that depend on each other to complete a capability (LEO or lunar). Still, our sparse experience here says we shouldn't expect to avoid a gap merely by saying we are mindful of the gap. Events after the loss of Columbia and the decision to end the Shuttle program led to the unexpected. For a post-ISS world, we have another opportunity to create the unexpected. This includes unexpectedly avoiding a gap as one program ends, and another builds and improves on what came before.

REVISITING THE NEAR FUTURE OF HUMAN SPACEFLIGHT

With acknowledgments to David Brin (blogging at Contrary Brin*) and a thank you for his feedback as I wrote this.*

Across my many years witnessing and participating in space marvels, too often my awe of the moment got rudely shoved aside by wondering what comes next. Some just can't resist that temptation to look ahead, to speculate in writing, and to re-evaluate past predictions. Lately, I perused some that physicist and science fiction author David Brin ventured in 2012 about the near future of human spaceflight. As we do in NASA decadal surveys, let's see how a look-back scores Brin's look-ahead.

In 2012, Space Shuttle orbiters were parading through the streets on the way to museums, after their last flight a year earlier, though we never imagined a Shuttle would also hold up traffic in LA. Right along, SpaceX first docked its Dragon spacecraft, delivering cargo to the International Space Station. Further behind, but still on our minds, were Spaceship One's historic crewed flights to the edge of Low Earth Orbit. So surely

Spaceship Two's offering for space tourists must be around the corner. Daring announcements were the norm, so Planetary Resources was going to mine asteroids, inflatable structures were in, and space hotels would be right along.

Back in NASA, which, contrary to the popular notion, did not shut down with the end of the Shuttle program, the course was firmly set for its new rocket -- the entirely expendable Space Launch System, built from Shuttle tooling and hardware, set to launch by 2016.

Unsurprisingly, what was coming didn't turn out quite as the headlines advertised. Dragon flights for crew would come, but they would take some time – starting eight years after that first cargo delivery. The billionaire Branson and Spaceship Two would eventually fly too, but only in 2021, and grounded since. For NASA's new rocket, "for sure in 2016" became the next year, over and over, until finally launching in late 2022. As for asteroid mining or space hotels, no, we are not there yet. It will take a lot more preparation and new waves of zealous investors for that frontier to look realistic.

What is surprising, that grand dreams happened much later than advertised, or not at all? Welcome to Space. What matters more is seeing what other marvels did happen, either on or off schedule. In 2012 we celebrated the successful landing of the Mars rover Curiosity. It's still exploring, almost 4,000 Martian Sols on Mars later! And a mighty dream – long deferred – was finally wrestled into reality by a different team of hugely competent techs and scientists: today we celebrate the similarly incredible, if complex and costly, James Webb Space Telescope.

Breathe in, breathe out. A day at a time, the Webb telescope may also be astounding us with its discoveries a decade from now. We have to ask, what else might be amazing us too?

* * *

Back in 2012, Brin said he was "inspired about our prospects in space." And there remain reasons to be inspired, if we take a close look at what's happening in the space sector. Perhaps because of this, I am told my writing is so optimistic, which is odd, as I spent my time in NASA developing a reputation as a bearer of bad news. "That won't be affordable. Here's the cost estimate." (Sticker shock.) "That's too complex and will spend most of its time in the hangar. Here are the numbers to back that up."

We knew how to pop technology balloons people floated long before it was in vogue. As time passed, I found myself in the company of others with these skills, quick to say "violates first law" or the classic "that's not going to work." These were skills, skills that made me a nightmare for project reviews.

Seeing rocks ahead as the boat crosses the river is more straightforward than it appears. The more real a project, the more pronounced the problems. This left ample time for me to realize that helping to find innovative, better paths forward was more productive – and enjoyable. Focus on the good news, and figuring out what's needed, and the optimism comes naturally.

Exploring possibilities – with faith in a bright future - is necessary for survival. (Or perhaps automatic, inevitable, and evolutionary? No matter, I'll take it.) NASA's budget may go up, but it's constantly losing purchasing power year after year. Simply put, NASA does not have the dollars, or time, to explore every great idea with R&D, a prototype, and a crater where a test stand once stood. Worse, if the dollars did show up, making craters is no longer in NASA's DNA (if exploring them still is.)

And yet the testing, prototyping, playing, blowing up, and failing must be done. How else to learn, eventually succeeding by doing? Of course, projects have problems, and problems quickly breed pessimism. All is not lost if we know we will try again. Since 2012 many more people have been exploring possibilities, not just NASA. Here are just a few projects that weren't on our visible horizons, ten years ago:

Want to explore 3D printing's potential to make more of your rocket quickly, in-house? NASA will explore some 3D-printed parts here and there, but why not 3D-print the whole rocket – like Relativity Space.

Believe the old notion of hurling a payload to space mechanically has reached a tipping point, now a mere matter of engineering? Then, let's try it out – at SpinLaunch.

Can we fuse structural, propulsion, and thermal management design so tightly it enables a reusable orbital rocket stage? While we are at it, let's do distributed propulsion – at Stoke Space.

Or perhaps it's time we bring in an automotive flair, remembering a rocket is a tank, with single-piston engines, minus the pistons. Can manufacturing yield economies of scale minus the fancy tech? Here, Astra is going old school with a factory on wheels, at scale.

And while NASA no longer says "reusable" much, it seems to be doing well for SpaceX, so let's build on that in new ways - says Rocket Lab.

And even more-speculative notions that are getting seed grants from NASA's Innovative & Advanced Concepts program – (NIAC) – such as new kinds of space optics, far side lunar radio telescopes, Titan submarines and new types of nuclear-cycle rockets, along with fungi-stabilized Mars habitats, Venus flyers, dozens of light-sail variants and neutrino-sensing missions to the Sun!

Then, of course, there is the elephant in the room – a Starship billed as a NASA lunar lander, where really NASA is investing in making space tankers. This means in-space refueling and the once-verboten space propellant depots, all based on a fully reusable launch vehicle. Everyone now, together, say "RLV."

Next door, we see advances in AI, with sizable private sector investment in quantum computing. The dollars flowing into fusion energy are no longer for updating the same old studies either – crossing into the hundreds of millions of dollars to be remembered next to producing net positive energy.

In another timeline, NASA's journey outward in our solar system would, like the Apollo program, have pushed all these rocket, software, computing, and power advances into existence by the brute force of necessity. Today, what we imagine as sci-fi may arrive in our timeline by other means.

※ ※ ※

The human spaceflight club is bound to expand its membership. NASA and SpaceX already went from an expendable launcher and a spacecraft carrying only cargo to reusing these, and carrying crew. Similarly, technology that seems far from human spaceflight today, orbital or sub-orbital, will easily benefit human spaceflight too. Failure (with cargo) will be an option because it's part of exploring, and how will we know otherwise – did we check all the possibilities? If we fail, not if but when, the matter will be how quickly we burnt through the list of what might work, and work best for the need. Once NASA was tasked to do this rather alone, now the private sector is allied in the cause.

The future will never turn out as expected. With NASA and industry investment together though, the millions here and billions there, the R&D with the let's go do it, it seems poised to turn out quite different. And it may be better than we might imagine.

THE NASA BUDGET - RUNNING IN PLACE OR GETTING AHEAD?

> *"My dear, here we must run as fast as we can, just to stay in place. And if you wish to go anywhere you must run twice as fast as that."*
>
> -THE RED QUEEN, IN ALICE IN WONDERLAND

Taking longer than initially planned or merely advertised is a hallmark of NASA projects, but the US Congress continues a similar tradition, usually delaying approving NASA's yearly budget. An astute observer is tempted to say delayed dollars for delayed projects – how appropriate! Then, in a now well-choreographed dance, projects blame some of their delay on the delay in their dollars showing up. Though maybe all we have here is an apt metaphor. Dollars or their projects are all in the same spaceship trying to reach orbit.

Congress approved this year's NASA budget just before the holidays in 2022, which might sound fine, except the Federal government begins its fiscal year on October 1st. On the bright side, better late than never. Also, Congress did increase NASA's budget by a generous 6 percent. That's not bad, but we also saw headlines about inflation last year at 8 percent. Though there's something to cheer about a 6 percent budget increase for 2023 when inflation appears on track to be on par. Close enough? If only being close did the trick. Unfortunately, like getting to orbit, close won't cut it.

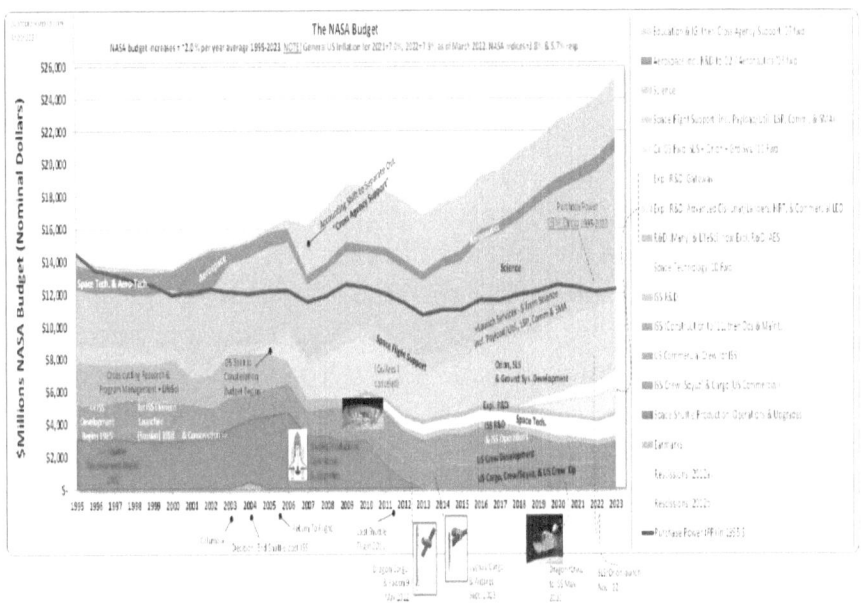

The NASA budget since 1995.

Since 1995, NASA's budget has declined by 15 percent - when adjusted for inflation. Slightly fewer dollars than would make up for inflation every year add up over a generation to a double-digit loss of purchase power. A scratch here, a small cut there, you feel faint from the blood loss before you know it. And this is a kind interpretation, that 15 percent loss of purchase power.

Can the yearly cost of science and spaceflight be tracked, like the price of a loaf of bread? Notice – there are two moving parts here, a cost, and the thing you buy, the bread. In practice, the patient experts who figure this out track similar aerospace products, but these are produced at a much higher volume than spaceflight. There is other technology, too, "close" enough to compare but not quite an identical loaf of bread.

It would not be surprising to find spaceflight, much like adjusting the budget of any other federal agency, simply costs as much more every year as is appropriated. The matter is what you purchase, rockets, satellites, planetary probes, and scientific knowledge. This is one complex basket of goods. What we put in our basket is weighed no more easily than adding up the bill. A common NASA saying about costs was *that if you like figuring out what a NASA project costs, you'll love figuring out the benefit.* Outside NASA, we hear a new phrase–"Shrinkflation," capturing the same sense. Did you pay the same for a smaller chocolate bar?

Last year and into 2023, the NASA inflation index crossed the number two compared to 1995. So, NASA now needs two dollars for every dollar it had in 1995 to get the equivalent (if it were easily measured) of science, benefits, hardware, or R&D. NASA's budget should be $29.9 billion this year, but it is $25.4 billion. That's the 15% loss of purchase power, according to NASA's own inflation indices.

Other measures are encouraging nonetheless. One of NASA's most important budget items, Cross Agency Support, which can be imagined as the fixed cost or "overhead" of any business, has been steadily dropping. Finally, someone said enough is enough and to make do. In 2012 this "support" cost consumed 21 percent of the yearly NASA budget, but it now stands at 15 percent. That's a clear indicator of improved efficiency – getting your overheads down.

At the pointy tip of the spear (or the rocket) and the business of going to space and back, NASA's commercial partnerships have increased from about 9 percent of the yearly budget to 17 percent in the last ten years. That would seem at odds with costs going down, but overall human spaceflight operations have dropped by about 8 percent since the Shuttle program's end (and that's before inflation.) It appears "going commercial" reduced NASA space operations costs overall.

Illustration by John Tennial

Needless to say, the year to start counting, or to compare, will give different answers. But mostly, these trends will hold. The real question asks, like the Red Queen in Alice in Wonderland, how fast does NASA have to run to stay in place, and how much faster must it go to get ahead?

Across perhaps as many years ahead as looking back here, we will learn if there is a real competitor to SpaceX in medium and heavy launch. Right now, nothing on the near-horizon appears worthy of note, in the US or elsewhere, from Europe to China to India. We see SpaceX rocking even small launch. Falcon 9 "Transporter" missions have removed hundreds of small payloads from the small launcher market. Transporter missions are like pesticides SpaceX sprays liberally, wiping out small launchers that might one day become bigger and challengers. Though Rocket Labs USA remains an ongoing, credible small launch provider, if the only one.

We can imagine a near future where we do know what

is in the basket of goods of "launch" reasonably well (if not as soon on the science, satellites, and probes side.) With NASA spending over half of its yearly spaceflight budget just on getting to orbit, where commercial launch goes, so goes NASA's ability to *deflate* a major cost. Yet in a chicken and egg dilemma, NASA's ability to encourage competition, fostering deflation, depends on its budget. The notion of innovation by desperation neglects the desperation needs resources, and latitude.

Not coincidentally, the rise of NASA's public-private partnerships encouraging non-NASA markets tracks with the decline of NASA's purchasing power and its renewed ambition to go beyond Earth orbit. Without launch and space system prices (costs to NASA) dropping, no budget scenarios for NASA to get beyond Earth orbit add up in reasonably relevant timeframes. Yet, as in design, it's important to distinguish a local minimum, a nice solution, versus the best. So, here's to this year's budget increase, providing some breathing room. But now is also a time, even when we have trouble measuring our stops and starts and value and dollars, to recognize time is always a wasting, and those dollars, and we all have a way to go.

A NASA IG REPORT, A STORY, AND A QUESTION FROM THE AUDIENCE

R ecently, the NASA Inspector General published another one of their periodic reports on the NASA projects that form NASA's effort to put people on the Moon. These IG reports are always insightful, if difficult reading. More than any other NASA organization, IG auditors have access to people and information in vast, complex NASA projects - and their task is to piece it all together for public consumption. I cannot help but recall a story or two as I digested the numbers and issues and parsed phrases for meaning.

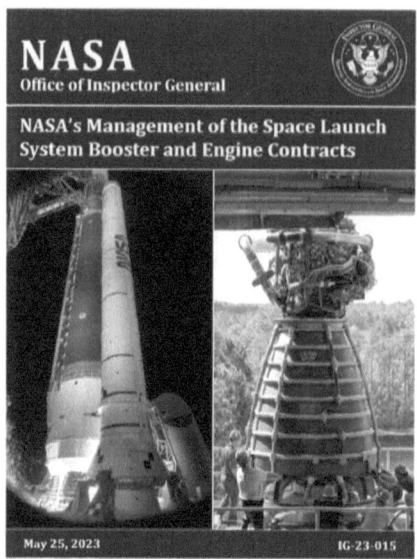

NASA Inspector General Report - NASA's Management of the Space Launch System Booster and Engine Contracts.

A full auditorium. The moving introduction is done with, the speeches too, something about safety and a special thank you to an employee who heroically helped on some job. As usual, it's always a thankless, unglamorous bureaucratic task. Thank you. On to the Q&A in the last five minutes, with apologies for (as always) running over time. A fearless audience member stands up (not me) and asks a question about direction, about it being problematic for obvious reasons, suggesting how we could do better. Standing up front, the captain does not rush to take the question. As in the movies, when the troops get impertinent, the sergeant in the front row stands up instead. Let's put this plainly. First of all, we have our orders. *You don't get to decide what we do here. This is what we have been funded to do.* Simple. The sudden shift in tone bought me back on site, my brain having tuned out earlier checking texts.

As happens with questions, and NASA, of course, when someone says, "Maybe this is a stupid question," they are immediately told in a pleasant, constructive voice, "There are no stupid questions." Perhaps the employee that day should have led with that. Or "I have an impertinent question." Instead, the stern reply said the quiet part out loud, along the lines of the teacher saying, "*You, come here, right now.*" Perhaps the director was having a bad day, perhaps the question could have been phrased in a manner more politic, or perhaps it was a way to say you can grumble all you want, just never in public. It may be the least memorable meeting leadership had all week. Just not to some of us straightening up in our chairs. The reply came across as scolding everyone in attendance, please listen up. (Best to quell the riot before it starts by making an example of the leader?) Or maybe, indicative of the problem, the answer confirmed the question that needed to be asked. Well, it was asked and answered.

The NASA Inspector General also asks impertinent questions. The recent questioning is especially arcane. Even for me. And I'm usually able to tell an interested reader where a number in an IG report comes from, with a spreadsheet. There are contract laws, award fees, regulations, and violations of regulations. The gist can be reduced, though – the IG says lots is amiss and problematic, then NASA replies, "You, come here, right now."

Of course, the IG is in an excellent spot to ask tough questions more than any employee. That doesn't mean the answers will be forthcoming. As in the auditorium that day, leadership may take the opportunity to remind the audience they are playing the cards they have been dealt. We have our orders. Next question.

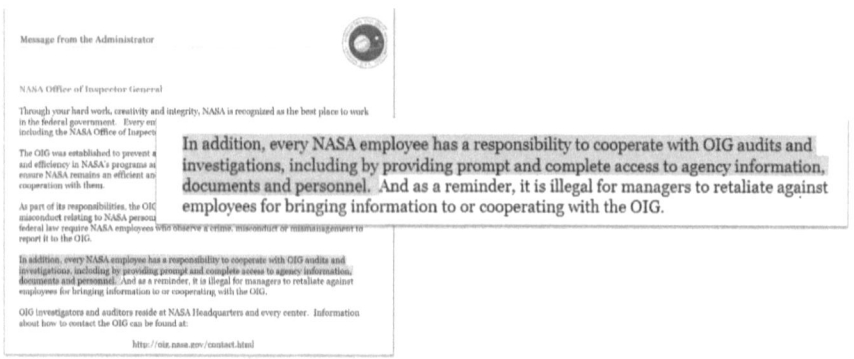

The yearly email reminder to NASA employees about their responsibilities to cooperate with NASA Inspector General audits.

NASA employees receive an email about the IG every year. It's a reminder to cooperate fully with any IG investigation. Employees must "cooperate with OIG audits and investigations, including by providing prompt and complete access to agency information, documents and personnel." This was an email to keep handy.

Supporting an audit is not one of those thankless jobs for which you will one day be recognized in the auditorium. I had the good fortune to be on both sides of the table, often the independent reviewer doing the asking, and sometimes the one trying to answer the tough questions. There are auditors outside of NASA, too, like the GAO. When they come knocking, you can see projects closing the curtains and making like they are not home. No one really looks forward to these fiscal colonoscopies. It's Monday, finally, that audit I've been looking forward to!

> **ARTEMIS PROGRAM MAKING PROGRESS BUT TIMETABLE SLIPPING BY MONTHS FOR TEST FLIGHTS AND YEARS FOR LUNAR LANDING**
>
> NASA's initial three Artemis missions – which are expected to culminate in a crewed lunar landing – face varying degrees of technical difficulties and delays that will push launch schedules from months to years past their current goals. With all necessary elements for the Artemis I mission now being integrated and tested at Kennedy, we estimate that NASA is progressing toward a launch by summer 2022 (a projected slip of about 6 months attributable to technical challenges, the COVID-19 pandemic, and multiple weather events – compared to the target date of November 2021). With Artemis II currently scheduled to launch in late 2023, NASA is facing longer schedule delays—until at least mid-2024—due to the second mission's reuse of Orion components from Artemis I. Finally, given the time needed to develop and fully test the HLS and NASA's next-generation spacesuits, the Agency will exceed its current timetable of landing humans on the Moon in late 2024 by several years.

NASA IG report November 2021.

Which brings us to this day when it's clear the lines between NASA projects and the watchers have broken down. This is not from seeing page after page about measly millions here and there. It's not about the "$5.6M unearned balance" or the (yawn...) $28.5M in award fees NASA seemed determined to give a contractor – when the law and all evidence indicated they should not. It's about the tone. It's that same tone I heard that day in the auditorium. "NASA leadership–was disappointed" and "did not concur" while the IG counters that NASA never provided "evidence to fundamentally change our findings and recommendations." Hint to projects and programs – bring your spreadsheets or forever hold your peace.

The root of this recent report lies elsewhere, back in the IG report of November 2021, receiving little fanfare at the time. Increasingly hinted at and danced around in reports stretching back years, we saw the real impertinent question. The IG concluded, "the Agency will exceed its current timetable of landing humans on the Moon in late 2024 *by several years.*" Alternatives are discussed, recommendations are provided. If it's measurable, it's good; if it's realistic, even better.

Yet here we arrive, at IG-23-015, the fifteenth report where the NASA IG looks at NASA projects and programs in the fiscal year 2023. Fees are amiss, engines are delayed, and NASA continues to reward poor performance no one is tracking anyway. Oh, and 2025 appears more like 2028. Maybe. But wait, if we might ask a two-part question, what with the recent Federal budget deal. If things looked this poor before, what will they look like when the budget in 2024 and 2025 only goes up one percent? Or maybe the NASA budget is frozen. Might we take a look at our options?

Time for everyone to stop checking texts and pay attention.

Advice I once heard for writers. Write like you lived it. My addition - if you did, even better.

CANCELED X-PLANES, CONTEXT, AND NASA

Context is everything. We easily commiserate with others when in similar poor straits or celebrate an achievement all the more in a backdrop of difficulties overcome. Either way, the best stories have scars in the scenery. Recently, NASA canceled another X-plane project, the electric X-57 aircraft. The usual NASA news with pictures from the James Webb space telescope, another launch, or a scolding of NASA's Moon plans was momentarily displaced, if only for one thirty-minute news cycle. Reload, headlines change, and new news. At this moment, we hear the all-electric aircraft begun in 2016 would not be; the technology is not yet there for this ambitious project. Having seen my share of canceled X-projects, from the hundreds of millions to the billions, the $134 million NASA will have spent on the X-57 seems par for the course. But I can't help but browse here or there outside NASA for some perspective.

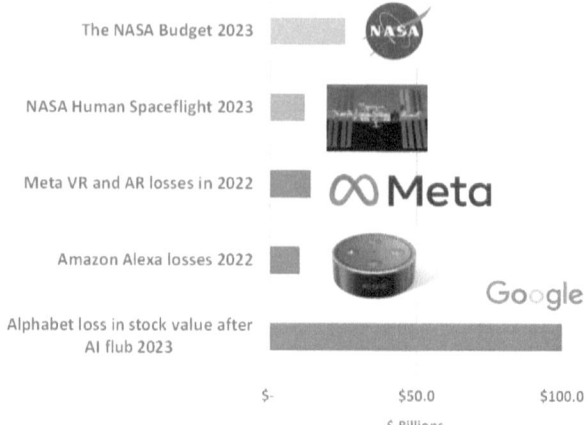

A billion here, a billion there, before you know it...

Earlier in 2023, Alphabet shares took a nose dive of $100 billion in a single day. Google's AI gave an incorrect answer to a question – about pictures from the NASA James Webb space telescope (so apropos) – and markets, meaning mostly large investors, took this as a bad sign. A NASA trivia question flusters the AI competition from Google and its sell, sell, sell. Poof, there went the budget of a small nation or two. For perspective, NASA's yearly budget in 2023 is just over $25 billion. And human spaceflight is less than half of that.

"Alexa, how much money did you lose Amazon in 2022?"

This is probably not the most popular question for Alexa – the answer is $10 billion, just shy of the entire NASA human spaceflight yearly budget. And by the way, rather than open the door and stick my head out, "Alexa, what's the weather like?" I expected a Hal 9000, but I got a way to re-order coffee without picking up my cell phone instead. (Which is still cool.)

There is a phrase from long ago in the sub-title of the 2009 Review of Human Spaceflight Plans Committee. It's "Seeking a Human Spaceflight Program Worthy of a Great Nation." At the time, the part *worthy of a great nation* drew my attention. Having supported the teams assisting the committee, conversations endlessly dissected the meaning of the phrase, those chicken bones strewn on the cave floor. Norm Augustine waxed on in interviews about the "worthy" part, seemingly frustrated. What's a few billion dollars in the grand scheme of things, in the context of such a great nation, and given the good NASA will put the money towards? That's the few billion floated as the extra funding to make life so much easier for NASA's grand plans to return to the Moon. What is the entire NASA budget, for that matter, popularized at about 10 cents a day per taxpayer? (My math comes to about 21 cents per person in the United States. It must be inflation.)

Over on the side of caution, context can also be damning, as when our mind plays tricks on us. In many analyses, providing advice, or reviewing the work of others, I was fond of hunting for biases. Selection bias, where we look for information that favors what we think is happening. We shrug off anything coming our way that runs counter to what we believe will be the correct answer. My mind's made up. Now let's go prove I was right all along. Mental accounting is another bias, where context warps the impact of a cost. We ignore that five-dollar bottle of water when we place it in the same bucket with tickets to Disney, literally a drop in an ocean of dollars.

Optimism is the most well-worn bias in NASA projects, and well documented, the context that consumes all the other ones. I am as susceptible to these biases (and optimism) as anyone, all the more reason to run "sanity checks" and play "devil's advocate" when our teams put out our answers. There is no vaccine yet against catching biases – again and again.

This will not be the last time we hear about a canceled NASA project. These projects begin with people walking into a room, *day one*. These were some of my favorite days. The air is full of hope, a fresh start, a new day, and I have that bounce in my step. But as Jeff Goldblum said in Jurassic Park II, "Ooh, ah,' that's how it always starts. But then later, there's running and screaming." Context works for and against us, sometimes the proper awe, the useful perspective, but also the obliviousness to what's ahead in the tunnel. I only need to remember which one is which.

"RESCUE PARTY," A SHORT STORY ABOUT NASA AND SPACEX – WRITTEN IN 1946

"Rescue Party" is a short story by Arthur C. Clarke written in 1946, but it is also the first time I read, at age eleven, about NASA and SpaceX (or "new space," generally.) That seems impossible, and off by about four decades, but it's true. Metaphorically speaking. The story is about time, the different speeds at which people move about the adventure and the challenge of growing and living. It's also a story about disruption and how complacency lies somewhere southwest of that. Without giving too much away on a sci-fi story that should be required reading for anyone trying to grasp the latest happenings in our space sector, "Rescue Party" is the culinary equivalent of new space meets old space. (You can read the entire short story right here.)

I came upon this story again, an old friend nearly forgotten on a shelf as I packed my books for a new destination and adventure. This was a gift from my 6th-grade teacher Mr. Brown – never one quick to smile, but always to be respected. He and Clarke gave me the gift of time travel, a wormhole connecting 1976 to 2023.

> As aliens explore an abandoned Earth in Arthur C. Clarke's "Rescue Party" –

> *"What do you think of this?" he said. "Suppose we've completely underestimated this people? Orostron did it once – he thought they could never have crossed space since they'd only known radio for two centuries. Hansur II told me that. Well, Orostron was all wrong."*

A NASA partner in crime (even in retirement) is fond of

saying, "We must often give up control for something better." Yet early in my career, this meme was not a going concern. Barely months had passed since a new cadre of troops were added to the ranks, myself included, and we youngsters were corralled into a small, musty conference room for a scolding. Back then, in dilapidated complexes that never saw better days, even when new, every conference room gave off a vibe between abandoned and neglected.

This day, someone got eager, was too obviously rolling their eyes, or did not "get it" about how things are done at NASA. There are well-established ways of doing things, and mentors must ensure new hires appreciate these ways (air quotes, or better yet, a mystery voice is a must when saying "ways of doing things," with emphasis on "ways.") The gentle scolding went on a while, between noting we were all so young and impatient. Frankly, all I took away from the lecture was someone had gotten everyone else in trouble. Who screwed up? Though admittedly, given the encyclopedic amount of unspoken tradition in a place like Kennedy, everyone seemed as likely to have set off the abundant tripwires, inevitably, including me. NASA Kennedy Space Center naturally impressed us "youngsters" just out of college, but the awesomeness wore off more quickly than you would expect.

At the year mark, a handful of employees in my hiring class decided they would leave NASA (a year being when you would not have to pay back moving expenses.) Why? "They said this is my job, then I found out no one would let me do it." She was not the only one. "The tile work on the Shuttles could be done much faster, but no one will listen." Later others joined in, and to an extent, the counter-notion flourished that "not everyone is a good fit for NASA." Good luck, and don't let the door... well, you know the rest. Except the farewells left lingering questions.

Besides excessive optimism, perhaps the worse bias burdening NASA's decision-making is the idea of sunk costs.

The "sunk cost" fallacy is a trick of the mind, failing to pick the right card by over-valuing our investment in the past. It could be any investment, whether what we paid or spent over many years or time and energy. We might even feel we invested in gathering "stuff"; having held on to something for so long, we must keep it a while longer. It's vintage!

Life may be the ultimate sunk cost fallacy. We are our past experiences and memories, or alternately "we" wander among the bookshelves filled with these, at arm's length, whoever that "self" is casually walking along. We value and over value the past even as what's done is done, inserting it into our calculations for what to do next. This is useful because remembering where the tiger almost got us last time, and the stillness in the air, is handy for increasing the odds the tiger will go hungry again. Yet economists remind us not to let the past delude us. Make your decision today, calculating only from now unto eternity or the relevant timeframe. We are warned to avoid the sunk cost fallacy, where we erroneously value the dollar already invested, irrelevant as it is to make a decision about the future. If you will make a million dollars, versus nothing, from today to ten years out, the calculation has no care for what you spent to date.

Decisions by the numbers are easier said than done. We are hardwired to look back, reminisce, and romantically value the past. Not so much on the looking forward with brutal honesty. Even as the world changes at warp factor five, we choose to go slow, holding on to tradition, the past, and "how things are done."

* * *

NASA now faces the prospect of a flat budget in coming years. It's time for hard decisions. Recently we learned the Mars Sample Return mission has a good chance of being the project that will "torch the whole science community." Here we are, *invested* already. Alternately, there is a possibility the project is canceled, unless it can show an honest prospect of staying within a $5.3 billion cost cap. NASA's plans to return to the Moon will also suffer, with the inevitable conversations around these monumental missions including some screaming about what NASA has already spent. How can we change course now? Will NASA, congress, and the Whitehouse seize the opportunities ahead, tacking into the wind, or will someone say we must stay *every* course, or worse, *every* detail – we've spent so much so far!

Twenty years from now, we'll know the answer.

NASA, SPACE PROJECTS AND CONTEXT – A MISSING LINK

Another day, another report by the Inspector General on NASA's big Moon rocket, the Space Launch System (SLS.) One day, we may see a new measure for project costs and duration of development, the number of IG, GAO, CBO, or other ABC agency reports. During a coffee break, someone in the crowd will say, "That project took eighteen IG reports," as everyone nods in sympathy, understanding the unit of measure.

RESULTS IN BRIEF

NASA's Transition of the Space Launch System to a Commercial Services Contract

October 12, 2023   IG-24-001 (A-23-08-00-HED)

WHY WE PERFORMED THIS AUDIT

The Artemis campaign seeks to return humans to the Moon's surface in 2025 before sending crewed missions to Mars in the 2030s. Key to this effort is development of the Space Launch System (SLS)—a two-stage, heavy-lift rocket that launches the Orion Multi-Purpose Crew Vehicle into space. In December 2022, Artemis I completed its 25-day uncrewed test mission after launch delays of nearly 4 years and billions of dollars in cost increases. NASA's total Artemis campaign costs are projected to reach $93 billion from fiscal year 2012 through 2025, with SLS Program costs representing 26 percent ($23.8 billion) of that total. NASA's development of space flight systems for Artemis IV includes the Gateway outpost, a Human Landing System, and a more powerful variant of the SLS rocket—known as the Block 1B—that will make the Artemis campaign more complicated and expensive.

In an effort to increase the affordability of Artemis, NASA is preparing to award a sole-sourced services contract, known as the Exploration Production and Operations Contract (EPOC), to Deep Space Transport, LLC (DST)—a newly formed joint venture of The Boeing Company and Northrop Grumman Systems Corporation—for the production, systems integration, and launch of at least 5 and up to 10 SLS flights beginning with Artemis V scheduled for 2029. Boeing and Northrop Grumman currently supply the SLS core and upper stages and boosters, respectively, that power the SLS. Before entering into EPOC, NASA intends to use a 3-year Pre-EPOC contract to evaluate DST's readiness to assume the new contract's tasks. Our audit projections estimate a single SLS rocket produced under EPOC will cost $2.5 billion, a figure NASA hopes to reduce by 50 percent through workforce reductions, manufacturing and contracting efficiencies, and expanding the SLS's user base. Given the enormous costs of the Artemis campaign, failure to achieve substantial savings will significantly hinder the sustainability of NASA's deep space human exploration efforts.

This report is another in a series of audits examining NASA's development of space flight systems for its Artemis IV and future missions. In this audit, we assessed the extent to which EPOC is positioned to achieve the Artemis campaign's performance and affordability goals. To complete this work, we interviewed officials with NASA, Boeing, The Aerospace Corporation, and the Defense Contract Management Agency (DCMA), as well as surveyed SLS Program and procurement officials. We also visited the Michoud Assembly Facility and reviewed contract files and other documentation related to SLS Program acquisitions, solicitations, costs, contract modifications, contractors, and production quality control.

NASA Inspector General, 2023.

    Usually, I read these reports seeking strange new words, new life in old recommendations, boldly going where no one has word searched before. Then comes the parsing, splitting hairs. Like a car crash, it's hard to turn away. Today, there is not much new here if you read the prior reports. Sequels, with rare exceptions, are not as good as the first.

    Even so, we have to talk. Something sticks out, as in nearly all of these reviews of complex projects mixed up in a stew of dreams and politics. Most of these reviews avoid context – the context being that NASA is looking at flat budgets ahead. But wait, if you order now, you will get more missing context – that this is no surprise.

    Arthur C. Clarke said there were three phases to a great idea:

1. It's completely impossible.
2. It's possible, but it's not worth doing.
3. I said it was a good idea all along!

What of the not-so-great ideas?

Clarke gives us one context for successful projects – the outsider's view. Success has many fathers, but not at first. But what of the projects that run into trouble? What of the *not-so-great* ideas? Decades ago, Department of Defense analyst A. Ernest Fitzgerald noted two phases to an increasing number of defense projects:

1. It's too early to tell.
2. It's too late to stop.

With only these two phases, imagine trying to get your project to right its course. At which stage are assess, re-plan, and re-orient? Fitzgerald saw how projects routinely turned their start-up phases, which should be open to change based on what's being learned, into years-long efforts reinforcing pre-mature decisions made on day one. This is not unique to the government. The private sector version is "fake it till you make it."

Yet this digresses from context, too – why bother to stop, change course, or improve? Why bother to be real? Why drive down costs (and price) while providing a better product, faster?

In recent budget news, Federal spending for most agencies, including NASA, will be frozen for a couple of years at least. In a year with just a few percent uptick in the price of labor and goods, a freeze amounts to a slight budget cut. Now, imagine our current years with inflation running much higher. Or, without needing to do math or imagine much at all, it's simply not if for lean budget years. It's when. For the private sector, the end of easy money is similar, now with high interest rates. It would have helped to aim low on your project budget expectations and high on your goals for what you will accomplish. But that's not how NASA's (or any agency's) project planning process works, building in ample margin for the rainy day and speeding up when skies are clear. Instead, projects plan for the best – the most unrealistic best. An old boss of mine had a saying when rumors stirred about tight times ahead, "We'll burn that bridge when we get to it." Obliviousness was policy.

On my shelf sits a book called "The Wrong Stuff," which tells the story of flight before (and after) the Wright brothers. "History has produced some amazing airplanes. And then there are these." Throughout, you will find innovations that seem like they might have been on to something if, for a moment, you forget what actually worked and why flight as we know it turned out as it did. If one wing provides lift, and two provide maneuverability, then nine wings must be better.

For all the science involved, managing research and development is not yet a science itself. Most R&D fails, somewhat by definition – if we mean by R&D things that are uncertain, difficult, and beset by "extraneous variables" – the last being a useful catchall for later saying, "It wasn't our fault." In Pharma, 90% of clinical drug development fails. The reasons for this are not clear. The circular semantics of getting pegged under "research" as if the label means "likely to fail" provides a license to wander. This is fine when the context – your local environment - is ample funding. Throw enough darts at the board, and eventually, one of them will hit dead center. That single success may pay for all the failures. But easy money is usually the exception, not the rule.

About a year before the NASA Constellation program was canceled, the Moon program before the current Artemis version, I was asked to give a big-picture analysis of its prospects. This may have been due to a particular skill. I could connect space technology, costs, and NASA budget scenarios and translate this all into something approaching English. Or maybe I was the only one dull enough to accept the task, not realizing no one wants to discuss what they can't control. (Remember those burnt bridges.) The scenarios were useful context – quantified. This backdrop was the focus, not the spaceships. What does the terrain look like, east or west, past the river or toward the hills? Needless to say, the prospects this first try at a Moon program would get the funding expected were less than nil. Or, more accurately, we entered the math of imaginary numbers. The program's financial executive responded with red-faced anger, though I was told my answers to questions were Vulcan-like. (It was my turn to be oblivious.) The message came across loud and clear. The fiscal environment, ISS end-of-life decisions, inflation, and cost trends were twelve decks above my station on the Enterprise. Though my analysis proved correct a year later, this is beside the point. Judging from the lack of such context in reviews since, it appears talking points about the context programs live in remains a faux pas.

"...commercial options"

Failure is fun when we pick up valuable, critical bits from the process. I participated in failed projects, canceled, ignored, or otherwise not selected at a rate that dwarfed the successes. Perhaps we have three phases to poor ideas that also lack situational awareness – as seen by project insiders.

1. Only this will work and cost the least.
2. We need lots more time and money.
3. You critics will get us canceled.

These phases have a common element: poor situational awareness. The final step is the most curious. Here, the news anchor Walter Cronkite is why the US lost the Vietnam War. One person, such power, to lose a war.

One day, the NASA IG or other critics may be blamed for losing the race to the Moon. Though the recent IG report includes some constructive advice on how to succeed

- "NASA may want to consider whether other commercial options should be a part of its mid-to long-term plans to support its ambitious space exploration goals." I can't help but sympathize with the challenge in messaging, as I often provided the unwanted advice. It was constructive advice, but only if you agreed the environment ahead mattered to your limited decisions today. Of course, this is not the first IG report suggesting a course adjustment; that was twelve IG reports ago.

> likely stymie any significant cost saving efforts. In addition, several FAR provisions may assist NASA in contract negotiations and mitigate the impact of schedule and cost overruns. Finally, in the long term, commercial competition in launch services will be more practicable for the Agency to better leverage less costly commercial alternatives while achieving its mission goals. Several U.S. space flight companies are already implementing multiple technological innovations, making heavy-lift systems lighter, cheaper, and reusable. In the end, failure to significantly reduce the high costs of the SLS launch vehicle will significantly hinder the overall sustainability of the Artemis campaign and NASA's deep space human exploration efforts.

"NASA's Transition of the Space Launch System to a Commercial Services Contract" NASA Inspector General, 2023.

THE ISS: FOR EVERY BEGINNING THERE IS AN END, OR NOT?

There are the facts, and then there is more. For the International Space Station, the facts are straightforward and measured, as NASA is wont to do. It begins long ago with endless concepts for a permanent foothold in space, move finally to the first hardware placed in orbit in 1998, a Russian module (purchased by the US government), and along to tragedy, the loss of Columbia and her crew in February 2003. Then a decision, the Space Shuttle flights will end, but not until completing the construction of the station (in twenty-two more flights.) Since the start, two hundred and seventy-three people from twenty-one nations have been to the station. Yet, with many years remaining and much can happen in the unforgiving environment of space, the ending to the station's story is not a given. Officially, as NASA plans go, the plan is to de-orbit the station around 2030. "De-orbit" is the technical version of saying NASA will bring the station to a fiery end, shoving it into Earth's atmosphere, causing a violent disassembly with its pieces scattered into the Pacific Ocean.

It's natural to believe this ending is written in stone, like the laws of physics, where all that goes up must come down, and all things age and decline. But can we imagine an alternate end to the ISS? Can we imagine no ending at all?

* * *

To temper our imagination, we need not go far. The sister program to the ISS was always the Space Shuttle. These two colossally complex systems overlapped in time, workforce, technology, and spirit. Their eras are more similar than different. With the last Shuttle flight in 2011, NASA moved on to the SpaceX Falcon 9 and other rockets as a service, leading to a newly vibrant US space sector merely hoped for at the time. SpaceX has since launched cargo to the ISS twenty-nine times and crew eight times. With the promise of space more compelling every day, private investors have put billions into other space companies as well. Leading the pack, the Falcon 9 rocket and the Dragon spacecraft carry on the torch of reusability once the sole domain of Shuttles. Falcons now launch every few days for customers from around the globe and to construct the massive Starlink constellation providing broadband services worldwide.

NASA astronaut Jessica Watkins conducting experiments on the International Space Station. Image: NASA.

This shift to NASA as an investor and advisor and to companies as partners has succeeded wildly. Beyond anything initially imagined, the end of the Shuttle story became a beginning for so much more.

This leads to a temptation. We don't need to overthink the end of the ISS, either why or what comes after. A mental jump powered by analogy says the end of the ISS will also be a new beginning for private sector growth we can't fully imagine.

Whereas the end of the Shuttle meant growth in getting "to" space, the sequel after the end of the ISS story will be about growth in the business of humans "in" space. NASA did this once. Now, rinse and repeat.

The analogy keeps on giving. Human spaceflight suffered a "gap" of nine years between the end of the Shuttle and the first crewed Dragon flight. Now, a "gap" repeats as a concern the ISS will be de-orbited, but no private commercial station will be ready to provide services to anyone, NASA or otherwise. The end of the Shuttle and the end of the ISS – similar plans, similar problems.

And yet, the notion of a station transition where NASA sings "Let it Go," moving on to being one customer among many, rests on the sense that there was a plan before for continuing to get to orbit. The plan succeeded, problems and all. Now, there is a similar plan for maintaining a human presence living and working in low Earth orbit. But is this only partly true, from a mix of myth, a belief NASA is a singular Borg-like entity and a misunderstanding of how NASA is built?

※ ※ ※

There was no shortage of negative opinion inside NASA as its "commercial" push began, where NASA partly funds partners, also asking them to develop non-NASA business. "After the cargo program fails, we'll have it as an example to shut up the crowd saying the private sector can do better. Here's the proof. We tried; they could not deliver." This story is recounted as relatively common by people in the commercial cargo program at the time. These critics did not expect SpaceX or other companies to succeed. If there was a plan, for some, it was for "new space" to fail.

"SpaceX prices are unsustainable. Any day now, they will renege on pricing or go bankrupt." This came from a senior NASA manager speaking with me at a conference. I began the chit-chat with curiosity about possibilities and a desire to understand differing scenarios. I was met with disdain as if the topic of cheap access to space were alien autopsy adjacent. More versions of this critique came my way later, uninvited, as I crunched the numbers for cargo. NASA benefitted enormously by getting supplies to the ISS with its newly available (but not owned) rockets and spacecraft versus the Shuttle. I was told that my numbers were correct by the contractual record, but the cheap pricing would prove a lie.

History repeated with the commercial crew program. "Boeing is the only one that can build a human-rated spacecraft." And, "Pressure is building to pick only one commercial crew provider – Boeing." And lastly, "What worked for cargo won't work for crew. Cargo is just dumb payload we don't really care about and can risk." The winds were shifting, and the commercial cargo program was no longer an easy target. So, a new narrative arose, where the success of cargo services didn't apply to what NASA did next.

Once again, the courageous efforts of the few fought the original plans - to fail, or await failure, to end competition, or to stop "commercial" initiatives after the cargo program. By 2011, I imagined a certain NASA manager in front of an oversized table covered with a map and toy models of rockets and spaceships. He emphatically strikes a spot *whack* with a wooden pointer stick, startling the underlings, saying, "We stop the commercial advance here!"

Predictably, success has many fathers (I was not one, busy

on another front), so this narrative arose where everything NASA did before to go commercial was always the plan. Not that a few didn't see it coming and pay a price to bring about success. Before the success of the commercial cargo program, there were advocates for NASA embracing an approach akin to how the FAA established demand for Air Mail to grow the civil aeronautics sector. Cargo to the station, the food, water, equipment, and experiments were the mail to deliver. None of these advocates for "making markets, not rockets" had it easy. (Then Deputy NASA Administrator Lori Garver abundantly documents this in her book "Escaping Gravity.")

With success, complicated events proving out a minority opinion unsurprisingly simplify. Soon, everyone says, "I said it was a good idea all along!" This is the real lesson when planning for what comes after the ISS. There are many motives and complex events in motion, and there is the advertised plan. Always look deeper to see if these are as connected as they seem.

※ ※ ※

The NASA plan for de-orbiting the ISS is part and parcel of a plan for a commercial continued human presence in orbit. Officially, this mashup gives you a straightforward doppelganger of NASA's prior shift around getting to space. After the commercial cargo and crew programs, another commercial approach has easily found its place in NASA. In a reversal of fortune, advocates for such "commercial" partnering would seem to have won the day. The war is over, "cost-plus" lost. Everyone really is on board with the plan, the plan is growth, and ending the ISS naturally follows. Or does it?

From the top - the goal is to grow our human presence in orbit to a degree we cannot imagine now. What follows? The connection between growth and the end of the ISS leads to a discussion about private sector investors, motives and NASA commitment. NASA must end the ISS as an operational capability if private investors are to believe NASA is truly committed to providing them business in the future. Investors must see NASA is closing the door to backups and other options, leaving no choice but to buy space and time on their future stations. Convincing commitment is a critical ingredient in NASA public-private partnerships.

Further, the logic goes that de-orbiting the ISS will free up funding NASA will use elsewhere. This fiscal factor is presented as a side-effect, a "plus" on the ledger rather than a driver. (As no one wants to say ending the ISS is merely to collect a windfall.)

Lastly, the justification goes, the ISS cannot be maintained in a "mothball" status, like so many NASA facilities awaiting some use. That could ruin the impression of commitment to future stations and zero out the budget savings NASA needs to pay for its next steps. There is no such thing as abandoning a facility in space, orbital decay meaning we must always pay to boost the thing occasionally, or gravity will take care of de-orbiting it for us. Uncontrollably.

It's a tidy package. This all seems proper, logical, and well-intentioned. Commitment. Fiscal reality. Moving on. Letting go. Some physics is thrown in, too, which, like DNA in a court case, is presented as irrefutable. Or are we missing something?

Unlike ending Shuttle flights, the ISS is already in orbit. Ending the station *operationally* does not require de-orbiting it. If we are using analogies to justify destroying the ISS, why not use the analogy to show why not. The Space Shuttles ended up in museums, why not the ISS – but in higher Earth orbit? As NASA ponders spending a billion dollars on a vehicle to push the station to a fiery end, it's worth at least as much to push it up instead of down. While no existing vehicle can boost the station to a "graveyard orbit," no current vehicle can de-orbit it either. A decommissioned station would be powered down, decompressed, and rendered immediately inoperable by lack of maintenance. NASA's commitment to future stations would remain clear, funding would free up, and a higher orbit would put the physics in our favor.

One day, two futures.

It is 2058. In the first timeline, NASA de-orbited the ISS in 2033 (delays in the plan, of course.) Somewhere in orbit a young woman is reading about the ISS, accompanied by a vague memory of childhood. The chapter is short. The wondrous ISS is nearly a footnote in history, as there is more to cover about goings on Earth's cis-lunar space. Business plans, materials, medicines, and math.

A second timeline. 2058. Here, NASA did not de-orbit the ISS. A tug approaches the ISS, one among the thousands of visitors a year. The sun illuminates the station as the tug circles around, the awe of an ancient cathedral at dawn. Inspiring, still performing its mission, the remains of the International Space Station effort point the way to the nations of Earth.

It was never only about zero-gravity experiments, studies or business plans. And back on Earth, for all that's been built up in orbit since, the ISS continues its mission, calling. Come, remember, see with your own eyes what's possible when we work together. And go further.

SLS AND ORION COSTS - THE THIRD RAIL OF NASA COST ESTIMATING

The usual conference room is crowded, with shiny surfaces and glass saying formal and stuffy, as I break in to ask my question about the numbers that seem way too low. What about "support"? Or re-phrasing, away from the specific and toward a broad sense, "The budgets have been much higher, so why is this so much less?"

"That's not included," is the reply. This classic answer is followed by silence.

The awkward pause comes as no one says those costs will be covered later to complete the picture or addressed anytime soon. Caution. No one wants to touch the third rail of cost estimating. Accompanying the cost estimate for the project will be a vague and barely noticeable disclaimer. *This is the cost estimate if we exclude most of the costs.* Those are the endless guidelines, assumptions, and obscure caveats for the project's numbers, translated here from NASA-speak.

If you are inclined to continue down the rabbit hole of what NASA projects cost, talking shop after the meeting leads to a confession. The estimator admits NASA's numbers do not follow – how do we say – "generally accepted accounting principles." Predictably, projects exclude as many costs as possible to look good. Later, to look even better, the project divides the initial cost estimate by pi for a list of reasons documented on one obscure chart in tiny print from three years earlier, slide fifty-seven, in the backup.

"We can't include all these costs, as they are *capabilities* NASA maintains regardless of projects." This is the end-state of these conversations in which I participated, often. I learned enough to be dangerous. Naturally, you jump to technology to improve a space system. To add context, you seek an understanding of the system the technology might go into. From there come justifications, improvements in performance, or costs.

Talking about support, capability, or the infamous "infrastructure" in NASA cost estimating leads to surreal moments. "That's not included" – is the response, over and over. You might be led to believe that new technology does not reduce costs, and neither do new ways of doing business. But what's not included in a cost when accounting for dollars does wonders. One trick is to get a project going as long as possible without *officially* starting. Many years of costs will later be excluded from the racking and stacking of funds spent. "That was pre-formulation. It's excluded." But it was the same contract, and it began the work? Again, an answer referring the questioner to another obscure chart with a diagram, arrows going in circles, and "guidelines" for the estimate.

Fair enough. If the cost is not here, it must be over somewhere. So long as we know the amounts, who cares about the labels? So, how much is it? "We can't show that, or it puts a target out there." A target? "A target when someone goes looking for money due to a budget shortfall."

This back-and-forth on NASA cost estimates is where inquisitiveness goes to die. It is a line of questioning that does not endear the inquirer to the presenter or project management. Conflicts of interest abound, as next time the analyst that did cover all the costs finds out management is getting another answer - from a different analyst.

These conversations are not the most surreal I have experienced in the years I spent estimating program, project, and technology costs. These are the relatively tame stories from the land of the costs. They are telling but only touch the surface of more significant challenges. For a Dali-level of surreal, melting clocks and all, you must jump to a word salad that should draw respect in the annals of word saladry. In 2017, replying to more impertinent questions about costs, an answer came back. Even for me, this one was new. It sounds so reasonable in a world of quaint blurbs about a bias for action. We hear this project is organized for results, not to make life easy for bean counters. If you want, "costs must be derived." Then, we discover the project's position is they are not required to derive those costs. (Laugh track in the background.)

Yet, because there are NASA budgets and public websites that track NASA contract data, and curiosity and perseverance aren't only rovers on Mars, a picture of the cost of NASA's major projects can be - *derived*. For the NASA SLS rocket, the Orion spacecraft, and ground systems, these numbers are consistent with other's totals. (Also with the NASA IG and GAO.) The new magic trick is figuring out the pieces one level (and more) below the totals. Once you have the pieces, "I just stick'em together."

> "I farm bits and pieces out to the guys who are much more brilliant than I am. I say, "Build me a laser" this. "Design me a molecular analyzer" that. They do, and I just stick 'em together."
>
> -JEFF GOLDBLUM (SETH BRUNDLE) IN "THE FLY"

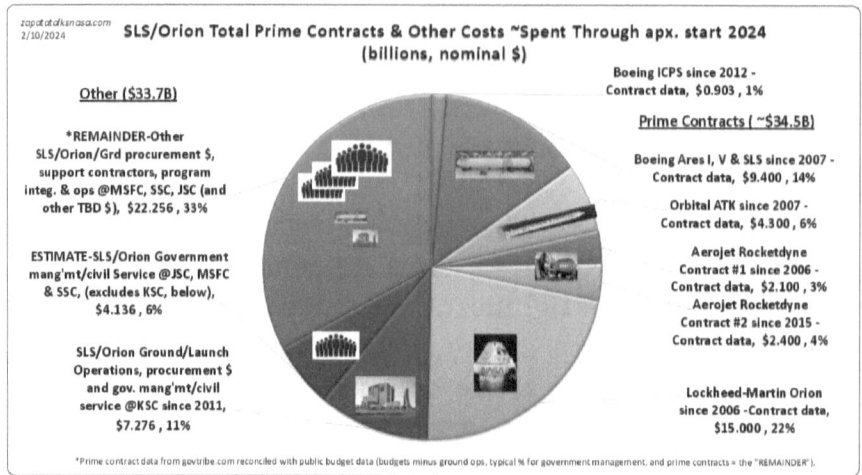

NASA's SLS, Orion, and ground systems have cost NASA about 68 billion dollars – to date. Prime contract data from govtribe.com. Total NASA budget data from NASA budget documents. All in nominal dollars.

In short, following the rule that Congress assigns funds to SLS, Orion, and ground systems, the "derived" parts and pieces come from knowing if it's not here, it must be over somewhere. Prime contracts get inordinate attention and cause confusion, even in reports by experts that assume readers know these are not all costs. (This context is usually overlooked.) Parse these all, and the remainder reveals the vast labor elsewhere at the different NASA centers with the parts and pieces. Prime. Ground. Civil servants. Capability and support. All gobbledygook, of course. But it's not a bad start at understanding costs, where, who, and for what.

In inflation-adjusted dollars, things get interesting. Adjusted for inflation to 2024, SLS/Orion costs to date are higher, about 85 billion dollars. This exceeds the development costs of the Space Shuttle, about 78 billion dollars, also inflation adjusted to 2024. Except the SLS, Orion and ground systems have a way to go to being declared operational. There are endless "puts and takes" analysts might reasonably debate here. Which years should be included under a project, vs. not as "pre-formulation" or ongoing work towards operational missions years away. These nuances will not change the broad strokes of SLS, Orion, or the Shuttle's development costs. Shifts in definitions will usually lower or raise both costs in lockstep. And for any debate, it's always worth remembering we can move on, toss some cost out of the development dollars, and put them over in operations. These are easier debates than they appear, if only for that reason - if it's not a cost to get something done (development), it must be a future cost after it's working (operational.) Trivially, if it's not here, it must be somewhere else.

The truly valuable debate occurs over value – cost by the mile traveled in space? Or to where? Learning how much? By the scientific value returned? Or in light of other options, in comparison by the pound, person-year in space, or the science per dollar. Time will tell.

There was another routine item in NASA cost estimating circles, a theme about anything Shuttle derived – "reduced costs through use of existing hardware and facilities." Perhaps, if the savings is not here, we can find it somewhere else.

* * *

Part 2

PART 2 REUSABILITY, SUSTAINABILITY, THOSE (BOTHERSOME) "-ILITIES" AND CHOICES

The astronaut Cooper (Matthew McConaughey) in the movie "Interstellar," revealing he will be staging from the main ship and leaving Brand (Hathaway) to go on alone, quotes Newton, "You gotta leave something behind." Action and reaction. A rocket climbing out of Earth's gravity well follows the same rule. To move forward, something must be left behind. It was never set in stone that this must be half the rocket, then another half, and more – like payload fairings and everything but the kitchen sink. Casually tossed aside into the ocean. Instead, propellant alone can do the job quite nicely. Staging never ruled out re-using the part left behind. As far as rocket equations and scientists do math, this is simple but not easy. The debate over throwing away the ship on the ride away from Earth, or in-space, seems settled. If we believe in the promise of SpaceX Starships to come, reuse is the future. The Falcon 9 already shows reusing *most* of the rocket is possible and makes sense. Now, it's for everyone else to decide if the debate is settled and if they will be around when the arguing is over.

REUSABILITY, PRICELESS.

There is a temptation to check off "sustainable" as a project feature merely because it appears likely to persist. Rather than this semi-circular definition, grappling with what is truly sustainable can move sideways. For one, sustainable space exploration and development can move to a measurable engineering feature - reusability. How much of something is reusable? What could be more sustainable than reusing what you already have and avoiding consuming resources to make a new one?

In my time at NASA, I saw reusability go from assumed to challenging to impossible and now (perhaps) to *necessary*. The impossible phase was after NASA's reusable launch vehicle programs in the late 1990s (X-33 days). Certain talk went around – the Space Shuttle, a lot like these new programs, also showed reuse was technically possible but prohibitively expensive. Not surprisingly, the expendable Constellation program would follow the loss of Columbia. Somehow, we had forgotten the talk that went around after Apollo (according to my co-workers in Apollo at the time), that massive expendable Saturn rockets, spacecraft, and lunar landers had proven technically possible but prohibitively expensive.

Yet now we are rediscovering reusability. How the world has changed.

NASA has indirectly embraced launcher reuse, with SpaceX

leading the way on the booster, not the part getting to orbit like the Space Shuttle. The first stage is reused now, not the last. SpaceX turned boosters around twice in 51 days in 2020, besting the record turnaround of 54 days on the Shuttle orbiter Atlantis in 1985. Just last week, SpaceX turned around another booster in *a mere 27 days*. Soon, SpaceX will use a "flight proven" booster to launch the 2nd US operational flight of a NASA crew to the ISS. This all happened more quickly than anyone expected. Even the yet to fly Orion spacecraft is being pushed toward reuse.

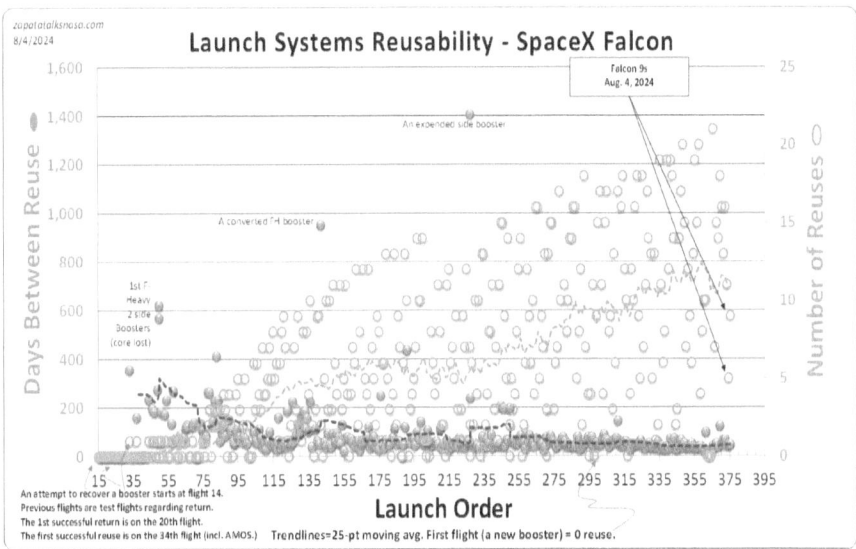

Days between reuse of the same booster and number of reuses for SpaceX reusable boosters.

This brings us to the older nay-saying on launcher reusability, where performance on a single flight was all that mattered. Never did a theoretical pound of payload in the future, ever figure out how to weigh so much more now, during design, development, or a single launch, as with a rocket.

Which brings up an obvious question. Is there any other industry on Earth where anything so expensive to manufacture as a rocket is one-use only? The tyranny of the rocket equation and the lack of experience may have limited us puny humans for a while, forcing us to ditch our ride as our tiny satellite barely made it to orbit. Decades later, as SpaceX is showing, the rocket equation hardly remains an excuse. The data shows steady progress on reusability – the number of reuses steadily increases, and the days between reusing the same booster steadily decrease.

Looking ahead, NASA's new lunar landers (in early design), at the pointy end of the spear, are also baking in significant degrees of reuse. As product lifespans tend to increase, so long as there is competition, it's natural to assume an economic eco-system where there is ongoing competition in space systems, in design, and throughout operations, whether in low Earth orbit or beyond to our Moon, will increasingly encourage more and more reuse. That reuse will make for space exploration that's ever more sustainable.

There's a saying about the three basic rules in life – never mix beer and wine, never fight a land war in Asia, and always reuse. If we are ever to go out to the stars and become "belters," drawing from near-endless resources beyond Earth, realizing the foolishness of cashing in the billion-year bonds back on

Earth, that 3rd rule should prove helpful. Then, how we got out there would be consistent with why we are going.

REUSABILITY - LEGS AND FINS OR WINGS AND THINGS?

The choice was made, so the outcome was determined, if not known. In engineering as in life. Not everyone accepts this notion quite the same way, or as gospel. Making a choice and then having to live with a determined if unknown future sounds fine in theory. In practice though determined leads to deterministic. As in destined. Fate. Meaning no longer in the realm of what can be affected or controlled. This is when the word "mitigate" comes in right on cue, because giving up control is not easy. In operations we were always on the tail end, facing the consequence of design decisions made long ago. The common refrain "sure we can" was part of the denial that the latitude in operations was thin, a reaction not a choice.

Design choices are like this. Something in the human mind wants it all. We want our choice (perhaps the easy one), but only with the very best outcome. Naturally. It's easier to believe a design choice made up-front and in the here and now will echo through the ages, when the outcome is grand. Not so much when the assessment comes back and says the future is not so rosy, because of decisions made months, even years ago.

Usually, this design rule is pictured where Y has to do with freedom, like the number of decisions already set (or the money already spent) and X is time. Your first steps decide so much. Destiny awaits, worse, the die has been cast and the actual form and shape of that destiny is still to be discovered. It's no wonder the concept was often touted about in any systems engineering slides and just as quickly set aside. A simple set of changes anytime ahead would still set the boat aright in the far-flung future-right? We could *mitigate* that. Actually, our design fate is what we make, but only if we make difficult decisions in the here and now.

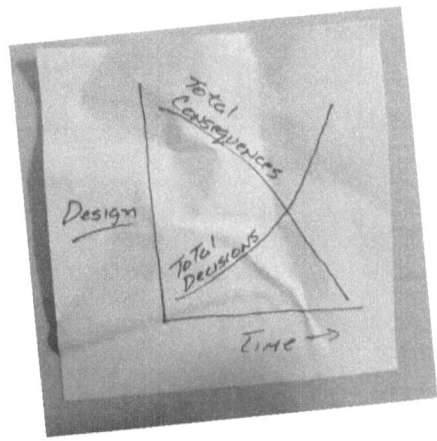

Choices, design and launch reusability

There are a particular set of launch system design choices around reusability. If we can think of examples where mass was missing, when it should have been there, being considered an enemy of the payload mass that is sacred, what happens when we think of adding mass so we can reuse a rocket? Better yet, to reuse a rocket *cheaply*? A first fork in the road is legs and fins or wings and things? Both designs can return a rocket stage to port. How to decide?

On first glance, each approach to returning a rocket booster would list pros and cons. For each, there are hidden items as well, things not as glaringly obvious as the fins or the flaps. Recently the DARPA XSP tried the wings and things on the booster, in contrast to the operational Falcon 9 booster returning thanks to legs and fins.

The Space Shuttle orbiter returning to the KSC runway, the Falcon 9 boosters returning to the landing zone after launch, and the envisioned DARPA XSP booster returning to the launch site after releasing its second stage.

The assumption choices also have clear outcomes to expect, results written in stone alongside the choices, must also assume more information. To the degree that information is absent, any destiny is less clear. It may still be just as inevitable and determined a result, just that it's not known. This we call risk.

Reusing a rocket via Horizontal Landing - Dissipate energy/velocity over the body, slowing down

Pros:
- Passive thermal protection does most of the work on return
- Smooth return
- Cross-range, can reach many landing runways very far apart

Cons:
- Complex thermal protection system to account for

cryogens plus aerodynamic heating
- Runways may not be where they are needed
- Active systems, actuated aero-control surfaces, landing gear

Reusing a Rocket via Vertical Landing - Dissipate energy/velocity by re-igniting engine, thrust to match descent

Pros:
- Landing sites can be mobile (a drone out at sea) or a smaller footprint on land
- Extensibility from expendable systems
- Extend learning to off-world systems (landers)

Cons:
- Must make the vehicle larger to hold extra propellant; propellant management
- Engine re-ignition close to ground taxes the engine and areas aft; aft wear and tear
- Active systems, actuated fins, landing legs
- Difficulty exiting landed vehicle in an emergency

There is a way to look at these two design choices that is not as stark as this or that – rather as decisions along a spectrum, that spectrum being the shape of the returning rocket. As measured by the ratio of length to diameter, go longer, skinnier and you have the Falcon 9. Add wings, like the XSP, or the SpaceX Starship, or the Space Shuttle Orbiters, and the ratio drops for the wider shape.

Seen this way, the choices for launcher reusability are on a spectrum. Curious, yet the conundrum persists – is there enough information that the measurable consequences of any of these choices are known? Probably not if we consider that reusable launch systems, weather the booster or the part that goes to orbit, are few and far between (TWO so far). Like the choices we make in our own lives, when there are few data points the consequences are not always clear (as much as we may have confidence otherwise).

* * *

Lacking information then, legs and fins or wings and things? Why not try each.

As much as seeing a Falcon 9 landing never gets old, as much as it may seem wings and things lead to extremely expensive Space Shuttle orbiters, we see in Starship mishaps that there remains much to learn about these design choices. Inside the extended barrel of a Falcon 9 (or Starship) comes things we don't see – for liquid acquisition. Pumps prefer liquid, not gas. Very predictable liquid. Similarly, behind the scenes for a winged shape are many surface actuators, gear doors and separation mechanisms.

We can't be sure yet of the real shape of things to come, shapes that may persist for generations, the way all airliners have looked alike since the 1960s. We can be sure there's a need for more public and private investment to figure out these choices and what they mean. Only then can we all make better choices ahead, with results still determined by choices, but also known ahead of time from practice in the real world.

X-33 – THE MIDDLE PATH?

I was walking under a beached whale, and inside it, and around, the dangling entrails smacking me in the face, an amateur mistake on my part. I should have known how to move carefully around flight hardware. It was early 1999 and the X-33 was taking shape. With its internal rib-like frame, and more platforms and curved supports and cables holding all the innards together, and the wiring like spaghetti all over, a beached whale is pretty much what it felt like. I had only been on these kinds of advanced projects a short while, part time, my day job still mostly in the Space Shuttle. The two worlds could not be further apart. Our team's task was to join them – the present and the future.

Much has been written about the X-33, its cause of death laid bare in many an autopsy, us experts standing around the carcass. Even ending up with parts of the carcass in a drawer. The project bit off more than it could chew, it had too many technologies not ready for prime time, not to mention the political ailments, all explaining its demise. All quite valid. But not enough has been said about why its design and its contractual and teaming arrangements were all so full of promise at the time.

An X-33 metallic tile. With this metal plate, anyone not paying attention to their surroundings would probably fare worse than the plate, quite the opposite of the delicate Space Shuttle surfaces.

If you are designing a whole reusable launch vehicle, or just part, a booster, an orbiter, a first fork in the road is the form *you believe* (at the time) follows from the function. *Here the function is to reuse.* If you go with a ratio of length to width that is high, you have a Falcon 9 booster. If you add wings and things and the ratio is lower you head toward shapes recalling the Space Shuttles. Early competitors to the X-33 did both, a vertical take-off, vertical landing on the one hand, a winged Shuttle shape on the other (revived by DARPA in the XSP program decades later). X-33 split the difference and went down the middle, originally at least.

The possible choices went across the design space, from vertical landing to horizontal landing, so from legs and reigniting engines to wings and coasting. Image: NASA.

Yet still not the blend expected.

Why add separate wings when you can still return with a graceful landing like an airplane by placing all your tanks and innards into a triangular shape that could do double duty for lift. The "lifting body" would be right up Lockheed's alley, from stealth technology and getting odd shapes to fly that look like they shouldn't. This was more than a design, Skunkworks would be an approach to how we would rethink process and practice, not just tanks, engines and tiles.

In this sense, the promise of X-33 was non-technical as much as technology. The non-technical points are often lost in the historical record, a contract form to more properly align incentives years ahead of the successful format that led to commercial cargo to the ISS. And the X-33 was for research, a non-operational demonstrator vehicle. On the technical side were things like robust metallic tile, eliminating an army of tile and blanket technicians as in Shuttle, and all their necessary support (the greater cost by far). Less advertised, differential throttling would eliminate large actuators for the engine, and where there were actuators, these would be electric and smaller, not a plumber's nightmare of hydraulics. A big one, the X-33 would not have toxic fuels either – an environmentally friendly step. The X-33's shape would do double-duty, eliminating large active aero-surfaces. *The body was the wing.* On gathering a team, the flatter organization, a firm fixed price, a bias for action, and a vision of mostly non-government customers once the full-scale ship was built all rounded out expectations.

In practice, the wings that began as small and stubby got bigger. So much for no wings and no large aerosurfaces with more actuators. The ship got heavier all over, and the path to a larger full scale orbital capable vehicle all in a single stage eventually seemed a bridge too far. Too late to matter, the immature technology of the composite tank that burst and sounded the end of the X-33 program was figured out, a few years later.

Too much of a good thing?

Forgotten in all this, individually, most early X-33 decisions did *lean* toward success. It was only later that it became apparent to most everyone that too many good choices were the equivalent of wanting it all, fighting a war on too many fronts vs. picking our battles. Many good decisions, it would seem, can end up not adding up any more than many bad ones. Horizontal landing? Vertical landing? Legs and fins or wings and things? Or a spot in the middle? As I look back, the best decision-making analysis and models at the time were isolated when sophisticated or barely connected when simple.

Decades later this does not have to be the case. While there is no replacement for simply doing, trial and error and experimental vehicles (and RUDs), the capability to understand decisions has come a long way thanks to know-how and capabilities we could only dream of in the late

1990s'. The temptation is to believe we could have made better decisions back then anyway, especially in retrospect, once all the information was gathered in one place. Yet we now have AI/artificial intelligence to help in search, finance, pattern recognition and knowing and predicting the shape of our own private lives (permission granted or not).

Perhaps one day the complex decisions we face with too many variables to count will all be mixed and combined and the AI will let us know, this set of technology, but not too much, for this result. If you are game. Do you want a reusable launch vehicle making a few flights every day like an airplane? Decisions, decisions. And that's a topic for more ahead – when we soon have to let the AI's do what they do well, help us with the design and the organizing people and the deciding. And we'll all enjoy the results of our good decisions come Monday morning.

YOU CAN'T ALWAYS GET WHAT YOU WANT, BUT...

The room filled with the usual suspects and small talk. This year it seemed an unwritten rule that before any presenter could talk about their good work there came this certain chart. It was the late 1990's, exciting times when ever faster computers, internet connections and aerospace technology came together to spur dreams of things to come. This chart was part of that. The future was simple, or so it seemed, a series of steps from today and getting something to Earth orbit costing $10,000 a pound to a fully reusable launcher costing just $1,000 a pound, and so on down the road. In a generation we will fly in machines merely 100 times more expensive than hopping aboard a flight to Atlanta. As expensive as that might seem it would be worlds cheaper than where we were.

Timeframe	Today	10 Years	25 Years	40 Years	Today
Launch Costs	$10,000/lb	$1,000/lb	$100/lb	$10/lb	$1/lb
Catastrophic Failure	1 in 200 Flights	1 in 10,000 Flights	1 in 1,000,000 Flights	1 in 1,000,000 Flights	1 in 2,000,000 Flights
Crew Escape	None	Yes	Yes	Not Required	Not Required
Fleet Flights Per Year	10	100	2,000	10,000	Millions
Turnaround Time	5 Months	1 Week	1 Day	2 Hours	1 Hour
People Required to Launch	170	10	2	None	None
Range Safety	Flight Unique	Mission Class Unique	Space Traffic Control	Aerospace Trafffic Control	Air Trafffic Control

Caggiano, Jackson, Muckstadt, Cornell University "Simulating Ground Support Capability for NASA's Reusable Launch Vehicle Program", 2001.

It was not quite so simple. The X-33, and next generation air-breathing vehicles launched from magnetic levitation catapults, among many technologies and systems NASA invested in, came and went. All the parts and pieces of hardware and studies never ended as flying machines. The costs would not drop just then, not yet, nor the prices atop those costs. At the time though, the lack of certainty on dropping the price of getting to space did not keep people from jumping to what they might pack as luggage once cheap rides became available. A special relationship between the ride to space and what goes inside the ride persists to this day.

One of the ideas back in the late 1990s for what might be the luggage for cheaper rides was Space Solar Power (SSP, or later 'SBSP' after a 2007 DOD study changed the name), a rebirth of an idea going back many decades. The champion for this idea was John Mankins at NASA Headquarters Office of Advanced Concepts and Technology. If NASA investments were to make getting to space cheaper, it would follow that the rides at these lower prices would also be increasingly for other than NASA and government agencies. Power stations in space were a natural next step.

Here in sunny Florida, a homeowner might get 5 to 6 hours a day of wonderful sunlight for their solar panels, and rather less on each side of that time as the sun rises and sets or when the weather is bad. Space Solar Power imagined a sun that never sets. Solar panels placed in Earth orbit would always face the sun and always transmit power to Earth 24 hours a day all year. This was not just a redirection of light, rather the power stations in space would transmit to Earth the power they received from sunlight after a conversion to microwaves. These were heady times when all this and more seemed possible. The power stations were massive, which is why they needed the transport cost to get down to that mere $100 a pound.

The vehicles however were not quite massive, but SSP would make up for that with a high launch rate. If a launch a week today makes for exciting times, the need here was for launches per day. It was on this side of the equation that a team at NASA Kennedy Space Center (myself included) were recruited to run the numbers. How many launches a day were possible given these technologies (if they became real that is)? Were the vehicle designs arguably consistent with launches per day? On what vehicles might this happen best, out of the assortment that seemed to grow every day, from pure rockets, to airbreathers, from larger to smaller, or single to two stage, but always reusable to be sustainable.

Mankins saw another link though beyond just a commercial opportunity and a world hungry for power, a demand to attach to a supply of vehicles. This was about more than just NASA moving toward being "one of many customers" for a ride (a phrase that did not become routine till decades later). Climate change was here and now, and what could NASA do about it?

The room had amphitheater seating, and the lights were dimmed during presentations. At our turn we had done our best to show how multiple launches a day were possible, to show the promise of how the far-away and impossible could be near and not so far-fetched. That is, *if* the designs from the vehicle teams had certain qualities, meaning technology advanced, mature and robust beyond anything at the time, *and* this led to drastically higher launch rates *and* drastically lower prices per launch, *then* building huge solar power stations in space would be possible. (This was a lot of "if's".) These would be affordable, operable launch vehicles, close cousins to the airliners that opened air travel to the general public, beyond the rich or the subsidized snail mail. For all the good work though, it was here in this dim setting as presenter after presenter chimed on that it dawned on me and others that all this was still *not good enough*. The dim setting was appropriate to the epiphany.

If we did not quite have the technology for the vehicle and the engines to add up to achieving orbit, we could help the takeoff with a few miles of advanced rail (think Battlestar Galactica-style). Credit: NASA.

The real problem was the demand for power on Earth grew faster than we could throw power towers in space at it. Even if what we could build made us look like an advanced alien race in its infrastructure heyday, it was far from what was needed. When I turn to Mankins, though, he is not surprised. Perhaps he already sensed the outcome. Adding up more detailed and better numbers across many experts and teams had not fundamentally changed the range on any of them from the start. Yet somehow, this philosophical resignation was matched by a desire to push on that persists today. Further progress would make the entire range move, eventually closing in on the target and adding up to power from space as fast as needed.

At the time, the numbers were staring at us all. Earth needs power, and even putting a massive solar power tower in space every year with thousands of tons of mass did not add up to keeping up with the growing power demand on Earth. For perspective, the most capable semi-reusable launch vehicle today, the Falcon Heavy, can put up about 35 tons at a time if its side boosters return to the launch site and its center booster lands on a drone ship at sea. Even if it might add up on launch frequency, that 35 tons today is still too expensive, as much headway as it's made to lower costs from where we were.

Of course, we could see a day when, with just more launches and lower prices, the power towers in space could keep up with Earthly demand for abundant, clean energy. The space tower technology could also get less expensive with time and more innovations like modules to ease manufacture on Earth. There would be experiments on Earth transmitting power across distances, such as the work of Nobuyuki Kaya in Japan. More recently, Paul Jaffe at the Naval Research Lab has focused on well-defined applications like military scenarios providing power to far-off outposts to reduce fuel convoys along dangerous supply lines. All of this boils down to that measure – the cost per pound (or kilogram). There is no lack of keeping track of this metric in the industry, not just for making concepts like space solar power possible but for a world of such concepts also waiting for their day in the sun.

* * *

Refueling, private stations, and more

In the decades since we envisioned what might be for SSP, there has been no lack of more construction ideas and travel destinations waiting in line for low fares. Pondering a cost per pound, an obvious question became a pound of what? It turns out most of what might end up in orbit to go anywhere beyond is just propellant. By 2011, a team at NASA was looking at refueling in space. Once again, as with SSP, if the fare for the ride was much lower to get to orbit, a distinct possibility from seeing where the SpaceX Falcon 9 was going, what might be possible with the Falcon Heavy on the drawing board? That task was undertaken by Charles Miller, NASA's Senior Advisor for Commercial Space at the time, who ran with the possibilities. Many launch vehicles could compete to launch lighter, near-empty spaceship stages, while others competed to load them up with propellant once in orbit. All this would make enormous sense at the lower prices per pound to low Earth orbit versus trying to get the mass up there all at once in any larger, more unique, less flown, and much more expensive launcher.

Miller and his team would report on this (myself again included) – only not everyone was convinced. Some would argue that refueling in space was immature, and in either case, Congress had already tasked NASA to build a giant rocket that could put up a large mass, propellant, and payload, all at once. The task did not end well for the team. The numbers favoring refueling were compelling then, and they remain so, but the timing for such change was off.

Persisting, 2015 offered a repeat performance. Now Miller was outside NASA leading an independent look, skipping the stop at a gas station in space, now with "tanker" stages directly off-loading propellant to the customer. Here, the gas station would come to you. The Evolvable Lunar Architecture study also included commercial lunar landers, perhaps two, perhaps one. All this would lead in phases to fully reusable landers refueling on the Moon from propellants mined and produced there. Using partnerships and going commercial, it would all add up if the promising recent history for this NASA investment approach were repeated. There could be a permanent and growing presence on the Moon much sooner and for much less NASA budget than anything envisioned.

The evolvable lunar architecture didn't need to de-orbit the ISS to free up funds to go to the Moon – a practical strength to the plan. The supply chain was strengthened along the way to the Moon rather than abandoning your rear at Earth orbit under the guise of limited resources. To Miller, it was "about a propellant economy in space," but not everyone got it. Again, the numbers were compelling and remain so, but the timing for such change was off. Only now, by not as much.

Fast forward to the middle of a global pandemic and awaken in 2021. Anyone who predicted such change would be accused of wishful thinking. In April, NASA's Perseverance rover has extracted oxygen on Mars. From the barely-there carbon-dioxide atmosphere, a small device called MOXIE has yanked out the oxygen and tossed aside the carbon in a process that might one day lead to refilling a ship on Mars with liquid oxygen. For refueling in space, NASA has awarded a contract to Eta Space to develop technology for transferring liquid oxygen. And atop all this, NASA has partnered with SpaceX to transfer 10 tons of liquid oxygen in Earth orbit. If all this sounds eerily again like waiting for low fares for that otherwise unaffordable trip, it is. Except the low fares no longer seem far away. Forgotten in this, liquid oxygen is liquid oxygen, and liquid methane is liquid methane, even if everyone's stages and what's packed atop those remain unique and less amenable to becoming commodities. The propellant remains most of the mass of getting anywhere, ready to take advantage of low costs per pound to orbit.

Also, all the refueling in orbit is now seriously back on the NASA table after a touch-and-go moment in early 2020, where NASA went commercial after all for its lunar lander. NASA's eventual commercial lander selection (pending a final outcome on challenges from the losing bidders) was a reusable approach – the SpaceX Starship. The timing for refueling, reusable landers, and landers as partnerships for future commercial services was no longer off. The day had come.

The list goes on for what's possible and the people who see the opportunities. As costs per pound drop, the commercialization of low Earth orbit is now on the table. This is also about more than just the launcher, as getting to low Earth orbit, as the saying goes, is halfway to anywhere in the solar system. It is no coincidence that a private mission to space will soon use the Dragon spacecraft developed for NASA crew trips to the International Space Station. The Inspiration4 mission – "four crew members representing the mission pillars of leadership, hope, generosity and prosperity" – would arguably not be going to space were it not for a vehicle and a spacecraft the reliability, safety, and price of which allow such a private mission. It helps that these are not government-owned or operated. Manufacturing can see a similar renaissance – as NASA's Lynn Harper, lead of integrative studies for the Space Portal Partnerships Office at NASA, is fond of saying – as we discover "where gravity is holding back US industry." Perhaps the breakthrough app for something made in space will be ultra-pure optical cable, ZBLAN at $2 million a kg, or a crystal used for a cancer drug (like Keytruda). It could as likely be something unexpected, more valuable than we can guess.

Refueling, the commercialization of low Earth orbit, from private space stations, for tourism, manufacturing, science, and research, helping us live longer, healthier lives, and yes, perhaps one day for space solar power is merely what we can imagine. The old "build it, and they will come" model may have been unambitious. In an era of thousands of Starlink satellites in orbit, what else might be on the table? Months ago, I had the rare opportunity to catch a recently launched fleet of Starlink satellites pass over the sky here in Orlando. A tidy row of bright lights streaming by on a clear evening from horizon to horizon leaves an impression. I wonder who might have dreamed a new dream on such an evening. For perspective, a completed Starlink constellation is of a mass of a multi-gigawatt Space Solar Power station – *and it's not waiting around for even lower costs on the ride and more often.* This could all be a fantastic picture of systems to get to space *and* systems and people in space that make the old NASA chart, with X-33, Maglev, and the laser-craft, look all too linear. It's enough to say that you don't always get what you want, but if you try sometimes, you get what you need.

An observer will be tempted to seek a common thread in all of this. It could be how innovative ideas are born, then wait for that "what-if" of lower costs to get to Earth's orbit. This is not a syndrome unique to the space industry, the ability to wonder what might be, say if a material were stronger or lighter, it's manufacturing cheaper. Our minds look ahead. What things we might do that day. The case for refueling in space, building power towers or private space stations, or advances from near-perfect fiber to life-saving medicines to scientific knowledge all depend on that cost metric. Yet, getting to know many of the innovators here over many years, the common thread seems elsewhere, in the perseverance of the champions of these possibilities before their time, without which the innovations wouldn't be quite as ready when the affordable rides arrive.

NOT IN OUR STARS

The saying "getting your wings clipped" took on a new meaning. It was 2002, and the plan for even a partially reusable replacement for the Space Shuttle now seemed a bridge too far. The debate had already devolved once, from fully reusable to partially reusable. It would devolve again, as all of a sudden, old-style capsules were no longer just the dull part of due diligence. Capsules were suddenly serious contenders. These were crew-carrying capsules, just like the Starliner Boeing will soon launch for NASA.

The Orbital Space Plane atop an expendable booster would have looked like a tiny Space Shuttle, wings and all. Yet amazingly, the smaller it got, the more it cost. Some might say the reverse, that getting costly the reaction was to make it smaller. The sub-compact must cost less than the SUV. More so, once NASA still gave everyone in Washington sticker shock, and as the newest policy circles turned their radar toward NASA, the billions to build a tiny Shuttle got the attention NASA didn't want. A little humor about the high cost of miniaturization would have been appropriate. Except no one was laughing.

Instead, there were the usual explanations, a favorite being a bit of Horton hears a Who, except where a function is a function, no matter how small. And all those functions cost the same regardless of scale. Why the new space plane didn't follow that logic in reverse and go big for a little additional cost never got proper debate. The Orbital Space Plane kept shrinking as if on auto-pilot till it seemed to disappear like the Incredible Shrinking Man. And so, we started to look down the path of capsules. Months later came the loss of Columbia, and capsules would soon prove to be the new favorites. Capsules were shapes amenable to adding emergency escape systems (for the way up) and simpler thermal protection systems (for the way down).

Fast forward to the NASA/Boeing Starliner that will soon undergo its second un-crewed sea trial (the first had its mishaps). But the week before that, a Starship also marked a milestone. NASA is partnering with SpaceX on a Starship as a lunar lander, and now award protests by competing partners have been put to bed. So, we might say we now have a NASA/SpaceX Starship that will be a lunar lander, or *a lunar lander larger than the Apollo Saturn V, and fully reusable*, which is a ship worthy of the name.

We have plenty of "Star" named things nowadays, after all. There is the Starliner, the Starships, and Starlink. But, like an iWhatever, the proper starter word is there to set a mood, if not about where we are, at least about where we want to go. (Starfleet, anyone?)

Progress, though, is a tricky thing to measure. All plenty of now you see it, now you don't. The debate over technology stagnation, as real or imagined, is as endless as perspectives on what's impressive vs. pedestrian. What do we expect in our lifetimes? A couple of features in technology discussions are missing when an optimist presents a case for being impressed.

Progress. We should count ourselves lucky. It's not every day you stand under a spaceship.

My wife, Arelis, beneath the Space Shuttle Discovery.

For one, it's the graphs. Find any work showing that technological advances have slowed down, and you can skip the reading and jump right to the charts. Measuring things that matter in the routine of people's daily lives - who would have thought. How fast are we moving, how long are we living, how different are our day-to-day lives? Is the robot finally for sale that walks the dog and folds the clothes or plots our demise? That would be progress.

Alternately, on the side of a slow march of progress, just hard to see, there are few if any graphs, but lots of inspiring stories. Two books by way of example - the pessimism in "Overshoot" comes from numbers and figures, while the optimism in "Abundance" comes from stories about people. "Look at this data," battles "I'm so excited." Where is the truth about the advance of technology, in data or anecdotes?

In between, we see where the future is always just around the corner, coming soon to a theater near you. The best example of this being Kurzweil, by sheer force of graphs wanting to show if we are not there yet, we soon will be.

Technology stagnation has graphs. Progress has stories.

There is another item behind more measures and graphs supporting the argument we are in a period of technology stagnation. As engineers and scientists, we tend to avoid economics even as much as it touches so dearly on the business of space. If progress is about remarkable changes defining how we look back on our lifetimes, then innovation is necessary, meaning competition. Yet, as economists point out, too many industries have all at once sorted themselves down to only one or two major players. This factor alone, the growth of vast monopolies, implies a period of technology stagnation. Where there is smoke, there is probably fire.

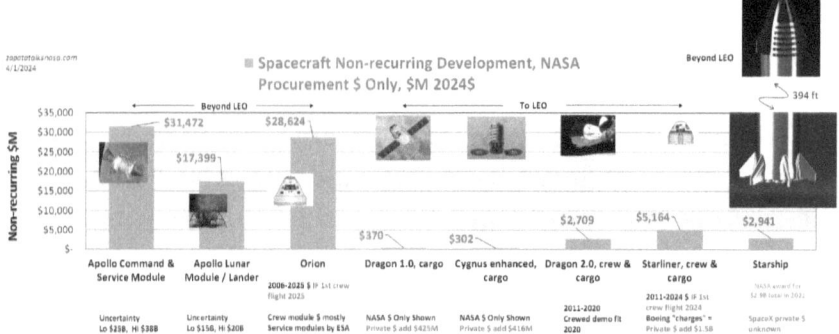

The up-front development cost of US spacecraft, cargo or crew, some as "cost-plus" contracting, others as "partnerships," some for low Earth orbit, others for beyond, some complete, some in progress. [All $ adjusted for inflation, as of April 2024-the Apollo crew and service module apx. $31.4B, the Apollo lunar lander apx. $17.4B, the Orion crew module $28.6B *to date*, the Dragon cargo spacecraft $370M, the Cygnus cargo spacecraft $302M, the Dragon crew spacecraft $2.7B, the Starliner crew spacecraft $5.2B *to date*, and the Starship lunar lander $2.9B *to date*.]

Yet what of our world of spaceships and spaceports? First, let's look at the numbers. It took billions of dollars to make a crew spacecraft of the capsule kind in the 1960s. It took about half as much to make a small lunar lander to match. Give or take, these numbers are telling (aside – the costs must also be set by which craft provided more push for the trip.) Today, Orion seems in the range of the older costs, as if there has been little progress. Although, arguably, Orion is a little bigger than the Apollo spaceship, and there weren't any opportunities to learn about building capsules and their modules after Apollo anyway.

On the other hand, cargo as far as low Earth orbit seems to have figured out cylinders and capsules, automated docking, and such, even reusable ones like Dragon. All for bargain-basement prices. Oddly, it would seem that adding crew to low Earth orbit adds a zero to the up-front costs of a cargo capsule/spaceship, with the few hundred million becoming the few billion. Perhaps this is not the result of adding life support to a cargo capsule so much as reducing risk (and adding Federal Acquisition Regulations, even if firm fixed price). Then, there is the NASA/Boeing Starliner, which followed on the heels of the NASA/SpaceX Dragon which has already taken US crew to the space station many times.

Which brings us to the other Star of the show. Having placed the Apollo lunar lander on the graph, we might as well place the lander selected for the Artemis program, the SpaceX Starship. If we can measure the lack of progress in numbers and graphs, what do we see here?

Offhand, this seems like a game where we are asked - *"which one of these does not belong."* Time will tell if Starships for a few billion to NASA are real or imagined. The graph will change, perhaps new data points here and there to give us a hint of what's to come. In the meantime, I heard they put twenty-nine engines on the Starship booster the other day. And they rolled to the launch pad the next. *Inspiring!*

October 13, 2024 update: SpaceX launched a fifth Starship test flight today. It also returned and captured the first stage booster in the launch pad's "mechazilla" mechanical arms. If we can measure a lack of progress in the numbers for some projects, we are also fortunate to see progress in the numbers for others, and in their fiery, inspiring sights too.

Credit: SpaceX

CONTRASTS

There are flying machines that stick in the mind, a Concorde, a Shuttle, a Valkyrie XB-70, or a Boeing 747. One machine that hardly flew but did this trick was the 1920s German Dornier X airliner, an early massive flying boat with 12 engines. Its wings had a crawl space so the crew could reach the engines and perform adjustments – in flight. Scary but true. The only one ever made was soon retired, so there was a lack of time to imprint on the public imagination. If the aircraft engine technology available at the time was not quite there yet, why, obviously, there had to be a way to – as we used to say in the Shuttle program – "mitigate" that. So, crawl space, engine adjustments – while flying. Yet right around the corner, a few years away, was the DC-3, a shiny, modern airplane, some of which are still flying to day.

For a few generations now, transportation technology seems to be at a standstill, and the turn in the curve and the DC-3-like leap have always been a mirage on the horizon. In NASA, I always knew a plan that did not reach its main goal before my interns would retire was not a plan. (Oddly, this was not the consensus as I was told, *"think long term."*) There has been no supersonic airliner after the Concorde. If anything, airspeed has slowed since the 1970s for the sake of gas mileage. Though speed is not the major hold-up in air travel just now, lines and security being what they are. We still await the bullet trains, too, at least in the US. And while we are at it, if any of this does arrive, it must be in a way that does not adversely affect the climate or Earth's limited resources.

Yet this last week would seem an exception, with signs of progress and urgency. There were engines, tiles, rollouts, and stacks that recall Shuttle engines, tiles, and stacks – except for the part about urgency.

Installing engines on a Space Shuttle orbiter was no quick affair. The first time I saw this first-hand, my senses fused into an oil-and-water combination of amazement and boredom. I was amazed at the precision and care of the technicians in physically demanding positions. I was also bored seeing so few pages of the paper procedure stamped off after so many hours. Patience is one of the particular set of skills I acquired in my years at NASA. Many years later, I would be looking at data for every engine ever installed at Kennedy, dates, tasks, teammates noticing that it *somewhat* matched replies engineers gave when asked about how long this or that task took. "*Somewhat*," meaning engineers invariably gave the low side of any time or effort when asked – an acquired skill for poor memories.

Engines were routinely removed from the Shuttle orbiters a few weeks after they were placed comfortably in the processing hanger. Platforms and protective covers were all over, and the hospital patient's IVs were all attached, as we said. Then, the engines were re-installed a month before the orbiter was ready to leave the hanger. By the mid-1990s, someone finally admitted on paper that removing engines between launches was unnecessary. Yet the task would continue, with good reason. There was so much work to do on the Shuttle orbiters that we had to remove the engines to make room for everyone else to work in the same area at the same time.

So, it was a pleasant surprise to wake up one day and see the SpaceX team installed twenty-nine engines – *in one night*. Twenty-nine. Someone will inevitably say these are slightly smaller engines than the Shuttle's, but not by much. In NASA-speak, the Shuttle engines were just the start of what we called our "plumber's nightmare," from the Shuttle orbiter all the way up and through to the orange external tank and out to the endless facilities. And that was just three engines. What the plumbing must be for twenty-nine engines on a Starship boggles the mind, mine at least.

A Shuttle engine hoisted up to be installed on a Space Shuttle orbiter in the Vehicle Assembly Building (VAB). This was not a regular event, as the engines were usually installed before the orbiter arrived in the VAB. NASA.

Tile was a more mysterious matter. When I arrived at Kennedy, our routine began with a morning gathering, coffee in hand, and a roundtable. Engines, check. Reaction control, check. So on and so on. A problem worth noting here, nothing worth reporting there, something off-topic to lighten things up. It was the comforting hum of everyone retelling the story about the day before. We were big on recounting events, and perhaps my writing here is what remains of that compulsion. The tile group had a calm and repetitious refrain, that moment in a church when you know the lines, something like "five tiles removed, three tiles replaced." And some new tiles with problems were added to the count. It was never much any single day. Some days, the count went backward, and more dents and dings were picked up than fixed.

The Shuttles "flying brickyard" is now the SpaceX Starship flying bricks, coolly hexagonal shaped. A generation or more removed, yes, and with what appears to be a novel means of attaching them, but still, the hexagonal bricks are a close cousin. Oddly, seeing a tile removed and replaced is an operation that as standard as it was for the Shuttle, is not one I ever saw first-hand. A tile would be missing one day, a hole in the pattern, and then sometime after, as if by magic, it was back, the technicians removing and replacing it in the middle of the night – between midnight and morning. It befell to do this when the least people were around. The noxious water-proofer we used was not something you wanted to have around hordes of people working nearby. The post-it note-like tags came some years into the Shuttle program, which Starship also uses. (As to urgency, we found out years into the program that bad tiles were only taken to the nearby shop once or twice a day in batches, not in the moment. And NASA did invent newer and much harder tiles, of course, but as the program ended, these had only made it onto the Shuttles at the rear.)

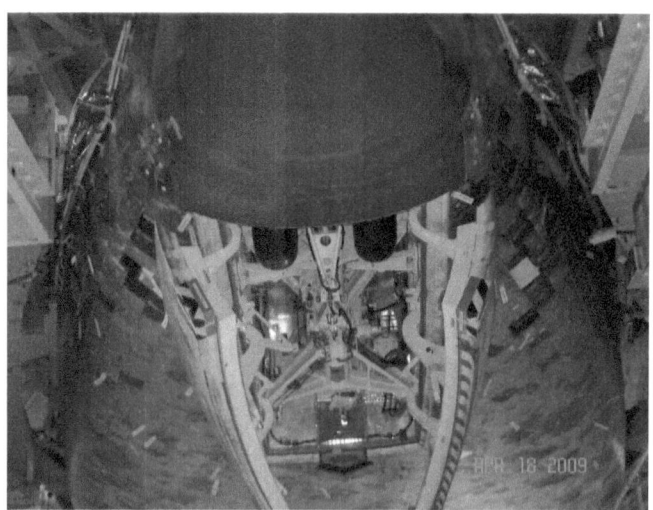

My photo of Discovery, front/nose gear, 2009.

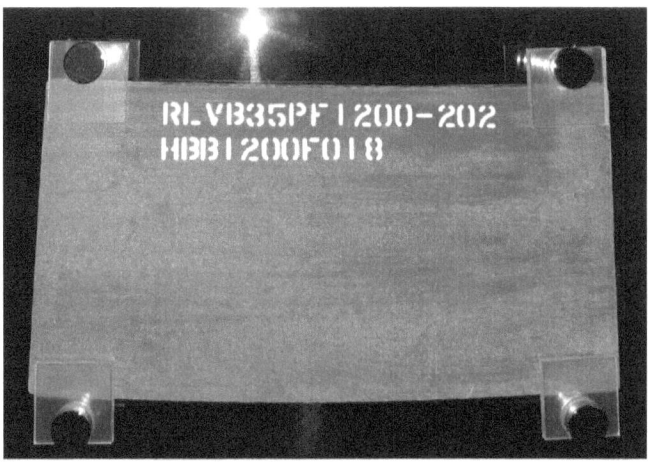

One the tiles we never got to install on the DARPA XSP.

The time to get thousands of square feet of tile ready for a flight is likely somewhere between the time for a Starship taking significant risks and a Shuttle taking none. That is, pending even better materials, an innovation around the corner that will put a skin on a spaceship coming back from orbit that needs nothing more than a good walkaround or some drone doing a surface scan.

But in all this, engines or tiles, something else is not as obvious in its absence. The Starship is not ensconced in a climate-controlled hangar. The SpaceX Starbase appears more shipyard than spaceyard. Having prepared and maintained liquid oxygen systems, flight and ground, I can say a firm job requirement is an obsession with cleanliness. We once had a liquid oxygen line open momentarily, outdoors, with some plastic flapping in the sea breeze to keep out the dust. Inevitably, there was a speck. And the tiniest speck that flew in would be wiped away, the rag with cleaner bagged (also noxious, eventually banned). UV lights were in vogue too once, leading to still more obsession over the cleanliness of flight hardware for liquid oxygen. (We discovered UV lights generate paperwork about residues.)

OPF Operations	Clothing Reqm't for ORB/HB's
Instruction NO 11 • Food, beverages, gum, tobacco products (cigarettes, cigars, snuff and chewing tobacco), matches and lighters ARE PROHIBITED in the OPF Hi-Bays. • Jewelry such as rings, watches, earings & necklaces shall be removed or taped securely to the body (athletic wrist bands are NOT ALLOWED), and glasses tethered when the individual is on any of the access platforms or in the Orbiter. • Lipstick, eye shadow and other makeup is NOT ALLOWED in the control areas. • Papers (WADS, prints, etc.) which are required in these areas shall be KEPT IN PLASTIC BAGS & removed only as necessary for reference. Papers shall not be held together by paper clips or "Desk Top" staples. **OPF Badging**	Bunny suits & booties shall be worn on levels 6,7,9,9A,13 & Orb P/L Bay & crew compt at all times. Booties shall be worn on levels 15 & and 21. Any request/chgs to the above shall be directed to LSOC OPF P/L Operations - 7-5475

From a 1988 Kennedy Space Center work schedule, a reminder about our clothing and attire while working around the Space Shuttles.

Obsessions of the Shuttle sort do not seem to be holding back the Starship spaceyard, and more likely, they need not anyway. The pressures and temperatures of rocket engine are not far removed from today's ever demanding aircraft engines - engines routinely removed and replaced quickly. The day too has come for the more advanced tile materials to shine, the next gen materials that never covered new Shuttles, and were only used here and there in Shuttles after seemingly endless analysis. As well, industry handles cryogenic systems quite well every day, minus the bunny suits and the obsessive cleanliness as a job requirement. Perhaps that sets us somewhere after the Dornier X, with some "mitigating" still to do, but at least on a straight line to some wondrous leap out of seemingly nowhere, like the DC-3.

As with the costs NASA and SpaceX are targeting for the Starship development, leaps might still happen – a hexagonal, mechanically attached tile, stainless steel weld, and twenty-nine engines at a time.

LIFE FINDS A WAY

NASA just rolled out a large, new rocket nearly eleven years after the last launch of its Space Shuttle. This is a long time coming, a project where too often "next year's" major milestones receded by about one and a half years every year. This expendable Shuttle-derived launch system will go down in history as what followed the semi-reusable Space Shuttle. For all the endless time and energy spent on studies, committees, arguing, and cost and technology analysis about expendable versus reusable rockets, the next NASA-exclusive launch system will be thrown away each launch. Yet is this the end of the story for reusable launch and NASA? I was told early in my career at NASA, and no doubt many heard something similar - "No" marks the start of the process. I took this saying to heart, in experience finding it true and helpful, more often than not.

> The Honorable John H. Marburger III
> Director, Office of Science and Technology Policy
> Executive Office of the President
> Washington, DC 20502
>
> Dear Dr. Marburger:
>
> In accordance with National Security Policy Directive 40, the Department of Defense (DoD) has coordinated on the space transportation strategy presented by the National Aeronautics and Space Administration (NASA).
>
> Recognizing the benefits of leveraging existing capability, as well as the cost and schedule burdens placed on unmanned payloads launched using human-rated systems, we understand that the DoD and NASA believe that separating human-rated space exploration from unmanned payload launch will best achieve reliable and affordable assured access to space while maintaining our industrial base in both liquid and solid propulsion launch systems.
>
> The major elements of the agreed strategy for the use and development of national launch systems are:
>
> 1. Both DoD and NASA will utilize the Evolved Expendable Launch Vehicle (EELV) for all intermediate and larger payloads for national security, civil, science, and International Space Station cargo re-supply missions in the 5-20 metric-ton-class to the maximum extent possible. As specified in NSPD-40, new commercially-developed launch capabilities will be allowed to compete for these missions if it becomes available.
>
> 2. NASA will initiate development of a Crew Launch Vehicle derived from Space Shuttle solid rocket boosters with a new upper-stage for human spaceflight missions in the 25-30 metric-ton-class following retirement of the Space Shuttle in 2010. NASA then plans to develop a new 100 metric-ton-class launch vehicle derived from existing capabilities with the Space Shuttle external tanks and solid rocket boosters for future missions to the Moon.
>
> 3. NASA and DoD will jointly pursue a cost-benefit analysis on phasing out the Delta II launch vehicle in favor of EELV to be completed in the coming months.

2005 Coordination memo, page 1, where then DOD Executive Agent for Space Sega and NASA Administrator Griffin divide up responsibilities in an expendable world. See Appendix B of the 2006 National Security Space Launch Report by RAND.

> Additional agreements relative to this topic are:
>
> 1. DoD will examine NASA's 100 metric-ton-class lift systems for any future DoD missions that might require such capability.
>
> 2. Pending a review of the final report of the Congressionally-directed RAND study into future DoD launch, it is unlikely DoD will endorse the use of a NASA-developed booster as a back-up for EELV due to the significant risk, reliability, and cost of modifications potentially required to DoD's satellites and infrastructure. NASA does not promote the use of a NASA-developed booster as a back-up for EELV.
>
> 3. DoD and NASA will commit to explore mutually beneficial cooperation for new upper stage development, advanced materials, other new propulsion technologies, and potential ride-shares on manned and unmanned missions.
>
> We look forward to working with you on the nation's space transportation strategy.
>
> Ronald M. Sega
> DoD Executive Agent For Space
>
> Michael D. Griffin
> NASA Administrator
>
> cc:
> NSC/Steve Hadley
> OMB/Josh Bolten
> DEPSECDEF/Gordon England
> OSD(I)/Dr. Cambone
> STRATCOM/General Cartwright, USMC
> SPACECOM/General Lord, USAF
> NRO/Dr. Kerr

2005 NASA and DOD coordination memo, page 2.

In ancient history (it seems), it's 1994, and the Whitehouse puts out the word. We operate a semi-reusable Space Shuttle. We have learned so much from that experience. NASA's next system will likely be even more reusable. Get to work. Then we can make "decisions on the development of next-generation reusable space transportation systems that greatly reduce the cost of access to space." Other U.S. Agencies, the Departments of Transportation and Commerce, would be "responsible for identifying and promoting innovative types of arrangements between the U.S. Government and the private sector." The Defense Department would continue to focus on expendable launch systems. It's easy to accuse the crystal ball of malfunctioning. Or maybe not.

By 2005 the tide turns. Reuse was out as the expendables

strike back. A new space policy was out, and the NASA Administrator coordinated the response with the Defense Department. NASA would go expendable, using "existing capabilities with the Space Shuttle external tanks and solid rocket boosters." The Defense Department did not change, holding steadfast since a decade earlier, still committed to expendable launchers.

This NASA sea-change happens in the wake of the loss of the Space Shuttle Columbia and the decision to use the Shuttle only to complete the construction of the International Space Station. As much as we learned with reusable Space Shuttle orbiters - Challenger, Columbia, Discovery, Atlantis, and Endeavour - at the cost of human lives, our next steps turned away from a long-term view. Where once we saw reuse as inevitable to have truly sustainable launch systems, now we would use what came before, the disposable parts anyway. We would continue using the Shuttle's solid rocket boosters, which failed Challenger. Boosters were at least recovered in the ocean, rebuilt, and refilled for reuse in the Shuttle's operations. The solid rocket boosters for the next system are entirely expended. We would also continue using tanks covered with foam, a large piece of which shed and hit Columbia, leading to another tragic loss. The only part of the Space Shuttle's lacking such failings, the reusable Shuttles themselves, would not inform a future system.

In a sense, this direction was inevitable. When it came time for advocacy about things reusable, any new system lacked a stakeholder with a pulse and a voice. Future reusable systems existed only in the abstract. On the other hand, existing expendable systems already had a seat at the table, complete with workforces raising their hands wildly like a student overly eager to answer the professor's question. One of many blind spots hid this reality, which any game theorist would have spotted much earlier.

And yet, expectations back in 1994 may not have been that far off the mark, if a little confused. "Innovative types of arrangements between the U.S. Government and the private sector" would be pursued, but not by the Transportation or Commerce Departments. Instead, NASA led the push for innovation with public-private partnerships for cargo to the ISS. This resulted in a new rocket, the SpaceX Falcon 9. Over time this rocket that began life fully expendable has become very reusable, with a first stage recently flying again for the twelfth time. Even the DOD would eventually reuse a rocket. This got done for the bargain-basement price of about a tenth of what it would have cost NASA under the usual contracting relationships endemic to old-style "by the hour" payment. (Guess... how ...long... that... takes.) Once flying, cargo to the ISS using Falcon and Antares launchers and Dragon and Cygnus spacecraft showed they not only cost less to create, they cost less to fly.

Illustration of SpaceX Starship human lander design that will carry the first NASA astronauts to the surface of the Moon under the Artemis program. Credit: SpaceX.

Similarly, a Starship is getting its "wet-dress" – the test of loading of propellants at the launch pad, just like NASA's SLS will do soon. Less advertised, NASA is a major investor in Starships - again as a partnership. The award to SpaceX in 2021 for its Starships as NASA Lunar Landers has a value of $2.89 billion. Consider that in a relationship between NASA and a company stressing results, paying for progress, not before or during, there is a significant factor of financial leverage. If one-tenth leverage has been floated (the high end of a range), that's $28.9 billion of result for just $2.89 billion of NASA's budget, more than has been spent to date on the SLS.

Among other news, in 2021 scientists confirmed that two California Condors had virgin births – Parthenogenesis. This fatherless creation is beyond rare in mammals, but to quote Goldberg, "life finds a way." As another reminder that life is much more interesting than we can ever predict, NASA actually has ended up leading the way on reusable launch. After all the work, hesitation, barriers, and tragedy, leaving one to think all those reusable eggs were infertile, some have unexpectedly hatched. NASA is learning to invest and provide expertise in new ways, while committing to be a future customer of reusable launch systems. There is a reusable Starship as a lunar lander within a NASA lunar exploration architecture, which has a pit-stop at a NASA Gateway in lunar orbit that will be supplied by reusable launchers.

Because after all, reusability finds a way.

ABOUT STARSHIPS, AND THE (NOT WHAT YOU THINK) REUSABILITY WE NEED

Recently, a SpaceX Starship ran into a setback that's been ongoing for a couple of years now – tile popping off. We get to see all this, as SpaceX runs a very open program, much of Starship taking place in the sights of a paparazzi of cameras and drones. We see that sticking protective tile on a spaceship is no trivial matter. Nor is this your parent's Space Shuttle or tiles. Then again, the Space Shuttle's iconic black tiles are too often over-simplified – they are a system of silicone adhesive, felt pads to allow movement, then tile. Even the best figures neglect to show more, like the adjacent gap fillers, preventing tiles from chattering and chipping against each other. The Shuttle flexed with the tremendous forces of launch, flight, and re-entry from orbit. For Starship, all the tiles will want to accommodate all that flex too, and worse.

A Starship takes Shuttle tile technology much further, placing them not atop an aluminum Shuttle structure, primed, green and clean, but rather atop a layer of insulation. Starships have their cold propellant inside them, not in a big orange External Tank like the Shuttles. This difference might sound trivial to anyone who ever put up some tile in a shower. But, except now for Starship, you might consider how you would stick tile directly onto that bathroom wall's insulation, minus the concrete backer board in-between.

An interested observer might wonder just what NASA did with all that research into reusable launch technology – billions of dollars' worth over decades. Long ago, NASA must have figured out this tile-on insulation nightmare. Oddly, the most telling public figure about how this *might* work is worse than the oversimplified figures of Space Shuttle tiles. A little googling for more in the public record won't hit the jackpot anytime soon. This leads to the question – is figuring out tile direct on cryogenic insulation a real technical challenge, or is something else going on?

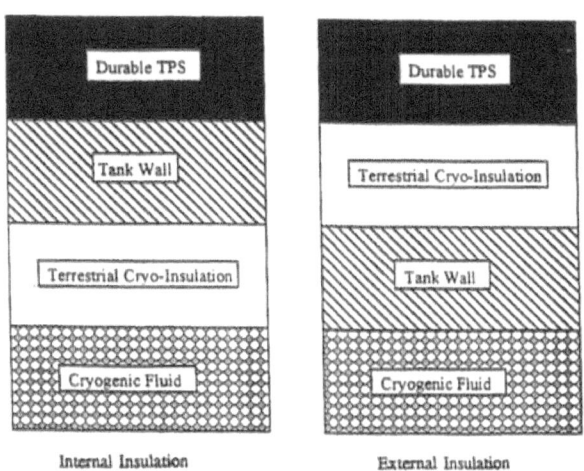

National Research Council, 1995, "Reusable Launch Vehicle: Technology Development and Test Program," Washington, DC, The National Academies Press.

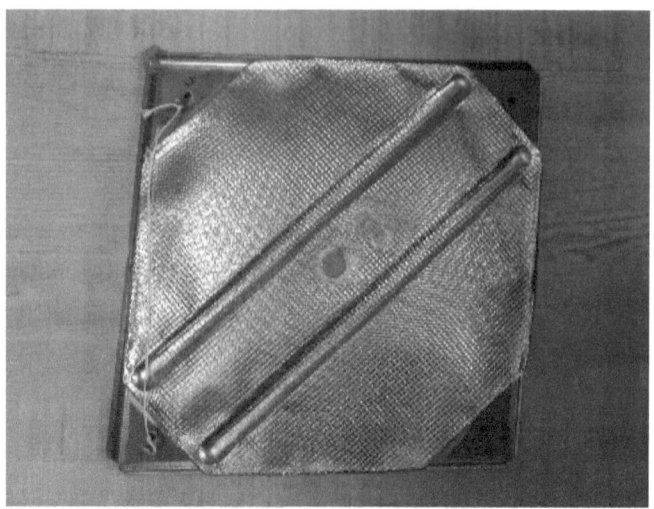

Rear view of a 2 X 2 feet metallic tile from the NASA X-33 program.

Kennedy Space Center in 1988 was a world of fantastic technology but not so incredible paper. Paper was everywhere. Paper filled our shelves, walls of cabinets, boxes under desks, and precariously asking for a fine from OSHA, boxes atop cabinets. And then there were the cabinets with combination locks. These held some sensitive information (/s) in the days when Soviet (Russian) trawlers awaited our launches offshore with bated breath just like the tourists. (Our secretary held the combination. Often, the room for these cabinets went unlocked after hours. Go figure.) But to really impress, Kennedy maintained a "secure" room in the Launch Control Center. Security de-bugged the room before any meeting, and the single phone line (right to Commissioner Gordon?) was supposedly super-secure (so we were told.) This was everyone's introduction to our duty to guard knowledge, loose lips, sinking ships, and all that.

Yet the cabinets or a secure room paled in comparison to the vault. Cabinets with combination locks and debugged meeting rooms came and went. We moved on, or the budget got cut, someone said. Yet well after 2000, this vault remained. Reading certain "Secret" material from the US DOD meant getting a hard copy delivered to this secure KSC room. Inside this government beige room, with too many coats of glossy paint, sat this large green bank vault. A little old lady managed access. She could be perfectly cast in a Quentin Tarantino version of a Mission Impossible movie. You could not take anything in or out of this room. I learned jokes here did not elicit a smile. I confirmed what I had already heard about DOD documents and over-classification.

Unsurprisingly, some technology needed work or progress meant a full-scale test next. All this was public knowledge. But if our knowledge loop felt constipated, technical or funding difficulties at these following steps only partly deserved blame.

Tile for a reusable spaceship with propellant tanks inside may have a long way to go. It is a system of tile, cryogenic insulation, fastening, or adhesives, with separation, height, and transition requirements. The holy grail, true reusability, will mean staying well below a max temperature. If even one return gets an area too hot, there are margins to spare, but that section will not be rated to fly again.

By 1999, the Space Shuttles thermal protection system was out there, documented in detail, and shared for all to see. Our next-generation systems – not so much. This is a reminder that if true launch vehicle reusability is a holy grail, *reusing knowledge* is also a challenge. Intellectual property and plain old competition explain why knowledge doesn't flow across companies. Regulations and caution explain other blockades, from the government to the private sector. And yet, time does overcome barriers, as Starliners or Dragon spacecraft re-learned most everything about thermal protection systems, and then some. Inevitably, we will figure out the technical barriers to launch reusability, sooner perhaps versus later. But it may be that knowledge that's reusable will take a bit longer.

ROCKETRY – IS IT MORE LIKE BAKING, OR COOKING?

Baking is not cooking, the same way rocketry is not flight. Or at least, that would be a first impression, to constantly hear about the extreme precision required to get to orbit or anywhere after. In contrast, right after takeoff, an airplane can lose an engine, or even both, only to glide along and land in the Hudson. The passengers end up safe and sound, with a story they will tell the grandchildren, over and over. Or take an explosive decompression. Aloha Airlines Flight 243 still made it to a runway, minus its upper front fuselage. It's assumed, flight can be forgiving in ways rocketry is not. Rocketry is different, like baking, requiring we follow a recipe precisely. Precise gets us to orbit. Close enough equals disastrous. And yet, is there more leeway in rocketry than we think?

Right now, two giant rockets will soon take their first test flights. The NASA Space Launch System completed (more or less) a wet dress rehearsal last week. Cryogenic propellants were fully loaded, as if mere seconds away from an actual liftoff. The precise sequence of tests began days before, but it pales compared to the strict requirements of everything in the minutes going to orbit. Also, just the week before, the SpaceX Starship obtained an approval (more or less) from the FAA for its first test flight. Unlike the first SLS rocket, SpaceX already built and tested multiple Starships, learning from each, including test flights of the Starship (upper portion), loading fuel and oxidizer, and firing up engines. One Starship even left the ground twice in the same day (sort of).

Here we have two rockets, trapped in the same physics, with wildly different ways to get to orbit. The SLS is an expendable rocket, throwing everything away to get its payload to a destination. The Starship will try to get even more delivered but return everything to Earth *spaceship included* to reuse next time.

The NASA Space Launch System. NASA.

The SpaceX Starship. Credit: SpaceX.

It's said that adding any math equation to a story means halving the potential audience. At the risk of losing some readers, it's worth putting "the rocket equation" here, if only to ask, what does this say in plain English? Admittedly, translating a famous equation (well, famous in my part of town) is not novel. It's common to see this version of naming a tune in as few notes as possible to get us thinking. (For example, Einstein's space-time and everything equation "$T' = T\sqrt{1-V^2/C^2}$" often goes under, "*the faster you move in space, the slower you move in time.*") Perhaps we can read the rocket equation "$\Delta v = u \ln(m_i/m)$" as telling us, "*The more you throw backward, the faster you move forward.*"

Staging, discarding, or separating part of your rocket, is always a great way to get rid of weight along the climb. Whatever push you provide after that point keeps you moving faster. The SLS takes this to heart, with large solid rocket boosters providing lots of push at the git-go, when the entire vehicle weighs the most it will ever weigh. Even more weight-efficient, these are soon discarded, this time *not* to be recovered as the Space Shuttle did. No parachutes for recovery, less weight to push along.

Starship, however, takes an entirely different route – figuring if you want your rocket back in one piece to use again, but you also want a usable payload, start by super-sizing the whole thing. SpaceX will also catch the Starship on its return, using a "mechazilla" contraption to avoid carrying legs or landing gear all the way to orbit and back. To top it off, SpaceX will refuel its Starships in Earth orbit, a way to get more stuff to throw out the back instead of only relying on what initial fuel you hauled up.

Spaceplanes, too, are more than familiar with this territory of tough decisions combining ingredients to satisfy the math. There are so many possible routes for a spaceplane putting a payload in orbit that even with the vast computing resources we now have, we still struggle to calculate the best path. A spaceplane is an initial decision - let's grab air along the way as part of what I throw out the back. A spaceplane's tough choice is deciding how much time it spends flying at around 100,000 to 150,000 feet, between about Mach 5 to 20. You want to grab as much air as possible to combine with fuel and hurl out the back so you move forward faster. Yet the longer you stay gathering air, the hotter it all gets, so you must carry more thermal protection. And all that free air fights you too, meaning more drag passing through that air, slowing you down. Now you end up spending more time accelerating, not yet ready to head up to enter Earth orbit. We haven't yet figured out a working spaceplane recipe, but there is no shortage of saying the measurements must be exact!

Rocketry seems very precise, with little to no room to maneuver, except most of the time, we see something very different. Raised on "the tyranny of the rocket equation," my fellow cult members once said reusability will never make sense. Then it was *full* reusability will never make sense. Some even said the complexity and difficulties of the Space Shuttles proved this to be the case. Why carry an entire spaceship to orbit, like the Shuttles, when it's just the crew or the cargo that needs to get there.

For the same reason, we are told spaceplanes will never make sense, even though one day we are debating how much maneuverability as airplane-like cross-range they might have.

You might land here or far over there. Airplane-like operations, flexibility, and much more margin for error don't make sense?

It's in the rocket equation. There are no ifs, ands, or buts about this, leaving almost zero room to imagine anything other than a retreat from the dream of reusable Shuttles. Voyages to Mars are also enamored with the rocket equation – if at times to close off possibilities. Ironically, we are now where hypersonics, which holds the promise of spaceplanes, is worked in the realm of weapons. This is happening while rockets, which look like missiles, move ahead in the civilian world.

So, the rocket equation is not as tyrannical as it seems. There is much more latitude for the imagination than we thought. We are free to at least try and cheat in the Kobayashi Maru test, because who likes to lose? Rocketry may be between baking and cooking, with latitude for a little of this, or more of that, after all.

SUSIE, SPACE LAUNCH, AND THE MANY JOURNEYS TO FULL REUSE

Over a week ago, Europe's ArianeGroup unveiled a new reusable launch vehicle they call "SUSIE," a "Smart Upper Stage for Innovative Exploration." Given the acronym, NASA must be rubbing off on them. Though the name is sure to be memorable, like Wall-E, reusable launcher announcements usually make a splash only to be quickly forgotten. But this is not just another startup putting out a press release that trends on Twitter while also saying they are in stealth mode. This is ArianeGroup, presenting at the International Astronautical Congress in Paris. This is when some (like me) in the space community say, "you had me at reuse."

Yet the details of SUSIE's funding should give us pause, and the bigger picture leaves curious questions. First, funding is unclear. You'll find the European Commission "Horizon Europe" program and €95.5 billion, which then, like Genesis, leads to a lot of begets. Horizon Europe begets space research, which begets more programs and acronyms, which begets reusability research, which begets some low 10s of millions of Euros for SUSIE. For anyone familiar with NASA's reusable launch projects, there is a little Deja-vu. An extensive portfolio begets this, then that, and many layers of the onion down, you find the relatively small projects where the rubber hits the road. In other words, SUSIE appears to be what we in NASA called study money.

SUSIE, Smart Upper Stage for Innovative Exploration. Credit: ArianeGroup.

While there is little funding, there is an immense technical challenge. SUSIE tackles the hardest part of reusability, bringing back the part of the rocket that goes all the way to orbit. A history of investments here could fill a book, starting with a Shuttle that put Orbiters the size of 737s in orbit and, near the end, delving into the SpaceX Falcon 9. SpaceX threw their design thinking into reverse, making the first stage reusable, not the last as with Shuttle.

Reusable launch investments are the bookends of my time in NASA's advanced projects. In X-33, circa the late 1990s, I played an insignificant role, asked a few times to provide an operational perspective. This was among the first advanced projects I was involved in, my full-time day job being over in the Shuttle program. A perk was my crawling around the X-33 structure once as it was taking shape. Lockheed would build liquid oxygen and hydrogen tanks, a launch facility, and a laundry list of parts and pieces.

Decades later, at the end of my career, I was heavily involved in the DARPA XSP program, one more charge at a reusable launch system. This program, too, built a liquid oxygen tank and another long list of parts and pieces, but with the launch pad and equipment left on the drawing board.

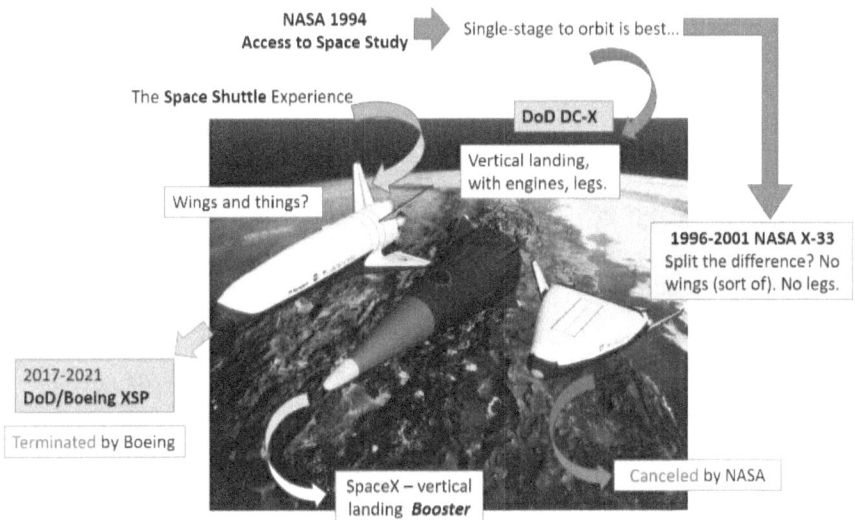

From studies like NASA's Access to Space Study, to programs and X-vehicles like the X-33, to eventual vertical landing and reuse and the SpaceX Falcon 9 booster, and another try at a winged reusable vehicle, the DARPA XSP. We are all connected.

Each of these programs met an early demise. As the X-33 ran into trouble, hydrogen tanks bursting counting as a bad day, Lockheed would find NASA ill inclined to provide more funding. Eventually, NASA canceled the X-33 program in 2001. Eerily similar, the DARPA XSP ran into trouble, this time with Boeing as a partner. Here, DARPA reminded Boeing this was a firm-fixed-price contract. Regardless of the scolding, Boeing opted to bow out of the partnership this time. One project was "canceled" by NASA, and the other "terminated" by the partner, Boeing. In theory, these are two very different beasts, canceling vs. termination. In practice, not so much. History repeated, the only difference being the time travelers were aware of the loop.

Design decisions in this space are, stating the obvious, complex. Yes, this is rocket science. Then there is SpaceX. They took an arguably successful design route, deciding that a reusable first stage or "booster" offered the most outstanding value. A pound of additional weight on a first stage does not mean a pound of payload lost. If you have too many miscalculations, too much optimism, or both, you have a good chance you might not stray too far from the payload you targeted. That final payload is also the market you want to serve and the promise you made to investors. The less that payload capability changes, and totally disappearing is possible, the better.

SUSIE will have the opposite problems, though these are not unsurmountable. As occurs for SpaceX with its Starship, being a pound off here or there on an orbital stage means a pound less useful payload delivered to orbit. These are complex challenges, but full reusability for sustainable, affordable systems must overcome these, eventually. The alternative is throwing away expensive hardware, the costs of which no aerospace manufacturing seen to date would seem to get around. Better sooner than later, with a full disclosure for the land mines ahead.

SUSIE and studies like it may at least make a small dent in a significant problem, as NASA did over the years. Perhaps one day, we will look back and wonder what all the fuss was about calculating a pound here or there or the technology we wrapped it in. Some studies here, some success there, a Starship, a SUSIE, or a Stokes, and before you know it, you might have full reuse.

SUSTAINABILITY AND NASA'S HUMAN SPACEFLIGHT PROGRAM: WE NEED TO TALK.

Throughout my career at NASA, I analyzed, prioritized, modeled, simulated, facilitated discussions and teams, wrote and reported, and got my hands dirty with, and crawled around lots of space technology. Eventually, I enlisted an AI-ish algorithm of sorts when it was clear some non-human help might do what a person or a team can't in a lifetime. But today, all I have is a story, and getting hung up on the meaning of some words.

It is so early that there is not even the initial sense of getting started when my contact from the other NASA center strolls in. The conference is called off. Slide decks sat ready, as did the coffee machine next to the neatly laid-out snacks. I made arrangements for tours because if you are at the Kennedy Space Center, I must remind our visitors there is a reality to what we are about to debate and decide. Still, we got stood up. The manager, many tiers above my pay grade, arrived at the center but sent his underling to say, "There's been some miscommunication." Having asked for ideas on projects that could save money, he meant save money – this year. Or save money now, yesterday preferably, from the tone as the conversation went back and forth. I tried to understand what happened. Ideas about future savings were of no interest, especially as it became clear the manager's initial financial liquidity had now evaporated.

Pick a card, pick any card.

This story dates to a few months before NASA's previous Moon program was canceled, about fourteen years ago. I vacated the conference room alone, calling down the list of presenters or greeting the early risers like myself with the bad news. Research and development and projects that advertised spend now, save later, were under fire for years, so the sudden order to halt today's festivities did not catch us off guard. We accepted that the artillery across the river would soon dial in on our coordinates. For a while now, the chatter was "Constellation program" and "unsustainable" in the same breath. The meeting's abrupt cancelation confirmed the diagnosis. Though, at the time, affordable, sustainable, and other close cousins remained words taking different and shifting shapes.

Imagine a factory making paper. Call it Dunder Mifflin, and you can easily see people talking about the paperless work environment, perhaps in a panic. Along comes a twist as the company discovers the word "sustainable," like the latest fashion it absolutely must buy. Some trees are planted, or at least it looks like that in the ad. Maybe it's endless trees. A tree logo. Slides about the importance of paper, sales still amazingly strong, even after email and "save a tree" campaigns. Everyone agrees that the business is infinitely sustainable.

This view of sustainability confuses importance and demand with supply and the long-term prospects for that supply. A view of sustainability as the agreed-upon significance of the task or the number of sales orders is incorrect and backward. Here, up becomes down, and down becomes up.

Now, years after that canceled conference to present ideas with future benefits and the schadenfreude soon after *everything* got canceled, it's worth asking what has changed in NASA's current Moon plans. In 2023, plenty of experts reviewed NASA's Moon exploration plans. The chatter is still what it has been for a while: *"Senior NASA officials told GAO that at current cost levels, the SLS program is unaffordable."* We also heard from the NASA Inspector General that the NASA SLS rocket has a price that is *"unsustainable."* In the polite version, NASA's current Moon plan has *"challenges."* In these reports, so many, umm, challenges. (A thesaurus would have been helpful.)

A tempting first reaction is redefining words – starting with "sustainable" and "affordable." These are a couple of classics in the hodgepodge of NASA-speak and the reports from the NASA Inspector General, the General Accountability Office, and the mattress mice (like me) who can't resist chiming in. In a Decadal Survey$_1$ we find that view of sustainability akin to stating the importance of paper or the likelihood that sales orders keep coming. When sustainable means *"widely accepted reasons to continue human lunar exploration that justify the continued investment, commitment, and risk*," we rightly see the importance of what NASA does and its goals. But we miss the mark, as missteps can still happen, even in important matters. We forget about the growth rate of the trees as if the paper plant exists in a vacuum. In this view, so long as demand persists, all is fine.

Another view on sustainability embraces our survival instinct. After the previous NASA Moon program was canceled, an amazing aspect is how the parts kept going, as if they never got the memo. Solid rocket boosters – check. Old Shuttle engines – check. The capsule for the crew – check. The giant rocket for it all – that, too, continued. Only an upper stage and its engine met their demise when the previous Moon program was forced into retreat. The empire, we found out, strikes back. And as for that upper-stage engine, it remained alive and kicking for many years after. There was persistence, continuity, staying power, and stability in the Shuttle hardware, infrastructure, and organizations.

I have roamed around the vestiges of the Apollo program that came before the Space Shuttle. In the 1990s, Kennedy Space Center still had the old massive Saturn V liquid oxygen pumps, purged with nitrogen 24/7/365, just in case. Years later, I explored the facility that now houses the Artemis program's Orion spacecraft, also built for Apollo. My initial foray into the facility found drips and leaks, mold and mice, and wishing I had dressed down that day. For the visual, think post-apocalyptic Hollywood B-movie. As the boss used to say, keep all the stuff you can when a program ends. One day, it gets us in the door to say we are ready for the next program. Keep a straight face on the bullet point "use of existing facilities reduces cost." Though frankly, it was about closing the door to anyone who said to do it right and from scratch or elsewhere. Once chosen for the next program, and after we upgrade the handle on the old hammer with a new one, and replace the head too, we would be ready to drill holes. With such proven persistence, it's easy to assume whatever NASA does is sustainable – on some level.

That's a pump.

The official NASA view on sustainability is close to what's typical outside our little aerospace world. Sustainability is about actions today that do not compromise tomorrow, creating the conditions to persist. Yet *"to execute NASA's mission without compromising our planet's resources so that future generations can meet their needs,"* says more about a factory conscious of waste. There is a long view here, but about the process, not the product out the gate.

A clear view of sustainability and human spaceflight can appear elusive. Seeing what we have is a start, avoiding the word salads and circular phrasing, sustainable as continuing to not be canceled. A critical mission – check. Clean execution, as conscious stewards of Earth's limited resources – check. Likely to continue to receive funding from Congress – check.

After importance, execution, and stakeholder support, questions still linger. Too many questions. We stood ready that day long ago, prepared to work toward a result far ahead. No mission, product or success is sustainable without looking ahead. Today, NASA's Artemis program at least exists in detente with NASA research and development (if an uneasy truce says the chatter.) No longer are there many voices saying every penny in NASA must divert to the objective of astronauts landing on the Moon. On another positive note, NASA went the way of commercial partnerships for its lunar landers and spacesuits – more than one for each. Not that NASA was given enough funds for any other path. Go commercial, and you can suddenly fund multiple landers for a few billion here and a few billion there where the previous plan couldn't see past one lander for tens of billions. That Starship slash lunar lander can be more, but a lander will do for now.

A decent R&D portfolio -check. An R&D portfolio that is no longer seen as merely a reserve fund for a lunar program – check. Respect for the proposition some investments only pay off in the very long term – check. Some commercial partnerships for significant elements of the Moon plan – check. To answer the question, a lot has changed since the last NASA moon program. Add these elements to the value of the endeavor, being good stewards along the way, and stakeholder support (mostly.) NASA scientists and engineers are fond of musing – this is all *necessary, but is it sufficient*?

R&D and commercial partnerships remain junior partners in the total NASA human spaceflight picture. A "sustainable NASA (lunar) Artemis program" screeches on the ears, over-valuing the stability of the means, not the achievement of an end. Still, continued means is necessary. A sustained human presence on the Moon soothes the nerves, remembering results, but risking the space station view of a home for a select few. That our solar system presence begins with a handful of people is natural and necessary. But bringing about a *growing* human presence in our solar system will inevitably need all this - then more. More R&D investment will be necessary to turn the curve, where more ambition is achieved for less than we can imagine today as we drown in some budget shortfalls. More commercial partnerships will also be necessary, as already proven, to fit ever more capabilities, enabling growth, in budgets that continue to lose purchasing power. Growth will not occur by doing less every year there are less (real) dollars. As to more ideas with some payback in the future, necessary, but it won't be about that either. It's about saving a future with a growing human spacefaring presence before word arrives it's been canceled.

THE NASA CIS-LUNAR UNIVERSE: LEO TO MARS+

The highlight of the sci-fi flick is the opening credits with the moody music, the point of view imagery as if floating in space, and the fly-through tour of the inside of a ship. Then it's downhill from there. This is what usually happens. Though this may be my peculiar view. I've been steeped too long in the space business. A suspension of disbelief is impossible when, five minutes in, I'm mumbling, "That's not how gravity works," or "There's no way that small thing has the propellant to do that." Ten minutes in, we stream something else. Is this why doctors can't enjoy medical dramas? "That's not how you hold that!" Once, talking to a fellow engineer who advised on a big-ticket movie, we laughed non-stop about the liftoff sequence, the dialog, and how the writers ignored his every input. The writers would say they know what works, the way Wile E. Coyote is funny precisely because he runs straight off a cliff, pauses mid-air to realize he has no ground beneath him, and only then falls. It's storytelling, not science class.

NASA is making its own movie now. The marquee franchise is the Artemis program to return astronauts to the Moon. The approach, or "architecture," is "Moon to Mars," as if to say the parts and pieces of the LEGO set. Think Marvel

Cinematic Universe, in which resides Captain America: The Winter Soldier, the best in the bunch, if you ask me. Though, of course, everyone's a critic.

Recently, NASA held a Moon to Mars workshop in Washington, DC. (Think focus groups for a new product or movie. What about the scent? What about the ending?) My recurring conversations leaned to the current production doing well, in contrast to the previous flop. (That, and what's up after retirement.) Though, my sampling also felt the first attempt at pulling together a Moon program years ago set the bar rather low. NASA's earlier lunar program somewhat met its demise in 2009. The pieces didn't connect too well, as far as physics. Worse, they did not fit inside the kind of NASA budget foreseeable by a sane person. Brinksmanship is not a space exploration strategy it turned out. Rocket, check. Spacecraft, check. No funding left over? Now, Congress must provide all the funds necessary for a lunar lander, right? Not quite. Though rather than give up, the "somewhat" part about the earlier lunar program's demise, most everything continued. The Space Shuttle-derived rocket and the crew spacecraft simply persisted without a means to land on the Moon.

Ten years later, having mulled over lunar landers for a while, NASA had new options. NASA awarded SpaceX a contract for a large lander in 2021, a variation of their Starship. Amazingly, this award came in at the budget indie film price of only a few billion dollars. Firm. And fixed price. The price allows many more pieces to follow – spacesuits, rovers (two types), and a Gateway station in lunar orbit to park it all. This was inevitable. With no change to business as usual, NASA would have a rocket, the Space Launch System, and a Crew Spacecraft, Orion, but no way to land on the Moon. If NASA started (or re-started) a business-as-usual lunar lander in 2021, judging from the experience to date with the rocket and the spacecraft, decades would go by. This kind of timeline arises when combining extremely high total costs with miserly yearly payments. As a rule, the probability of such an approach being overcome by events approaches one.

But NASA changed how it does business, after all. The re-booted NASA cis-lunar universe parts and pieces have not quite formed around some infinity stones, but there are signs the writers have a plan. Except, looking back on earlier franchise failings, Houston may still have a problem.

※ ※ ※

Cis-lunar space in National policy is the "region of space beyond Earth's geosynchronous orbit." Over in low Earth orbit (LEO), NASA plans to use private space stations after de-orbiting the International Space Station (ISS). This will allow "NASA to focus government resources on deep space exploration through the Artemis program." This is the quiet part written down - how plans for the Moon mean NASA budget dollars must be freed up elsewhere. It's "Moon to Mars," leaving LEO behind for others. The steps ahead are lunar orbit, then the Moon, then Mars. The connected steps, that is.

"You can't get there from here." This is not what anyone wants to hear when asking for directions. Though it's common when trying to cross the street to that restaurant you see from your hotel, in Florida. Something in the steps does not connect. In space, no one can hear your frustration.

Every dollar through any location in space includes the dollars to get past the previous point. Like rockets and "delta-v," where we measure changes in velocity, skipping steps is not allowed. It must all add up to "get there from here." The rhyme to this is intuitive to a military mindset in a campaign. The territory controlled behind your front line secures and enables your forward movement. Neglect your rear, and it may collapse. The supply chain will be unable to support your forward momentum. This lesson is fundamental in NASA and in DoD, but perhaps it is a habit to neglect it. This world is where large programs are rewarded for getting ahead of themselves and scaling up quickly before anyone notices. The survival strategy is to grow fast, then be too big to cancel. We go from an enthusiastic early phase lasting years, where everyone is looking at everything, to the next phase where, no matter what we found, "it's too late to stop."

Some sci-fi follows. Credits. Moody music. A starfield. An alternate universe. The fundamental connection from LEO to far beyond came to be called "LEO to Mars+." As the historian Nazzar noted in 2036, "National investments in an ever-expanding human presence in space evolved as they would inevitably. As a foremost example, the shape of the NASA budget in this golden decade reflected and took advantage of dramatic decreases in the cost of human activity in low Earth orbit. Only afterward did human activity in cis-lunar space begin to outgrow its earlier limited flirtation with a lunar presence." Human access to LEO is now routine, frequent, and cheap. Because of this, NASA funding (and Congressional interests) freed up to go further. Ironically, early attempts to speed up lunar exploration by not spending on LEO delayed further advance.

Continued NASA investments in low Earth orbit were critical. The notion of NASA as an "anchor tenant" buying time and space on private space stations initiated much more. The necessity of a vibrant orbital ecosystem spawned entirely new industries, from in-space medical research and production to semiconductors. In the same way the Apollo program in the 1960s needed technology that did not exist in order to reach the Moon, going on to invent it, the new low Earth orbit economy boom created industries that did not exist in order to reach even further, to stay.

Eventually, in this alt-NASA universe, the same happened in cis-lunar space. Human activity in cis-lunar space and the Moon became so commonplace that NASA moved on to Mars within its measly, pedestrian budget. The increasing investment in space by others, from the US Space Force, the Space Development Agency and the private sector rounded out an era of innovation. It finally added up, the only way it ever would.

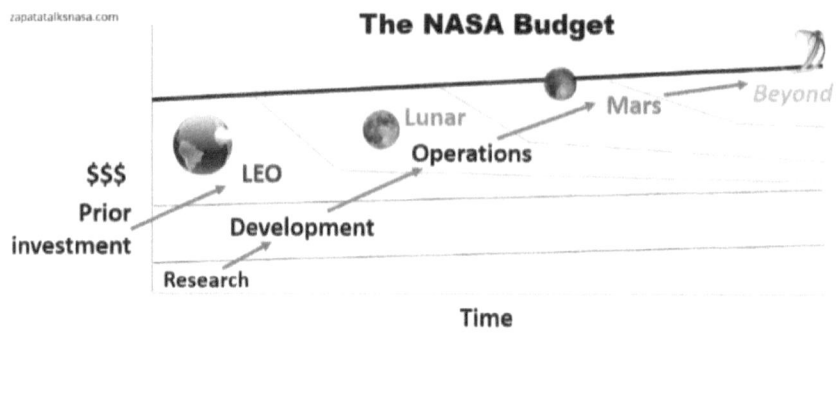

* * *

In NASA's earlier lunar program (called Constellation, a shame we can't use that tagline again), one glaring flaw arose from pushing energy needs away from the nearer projects in work. Energy. It's what it's all about. A rocket might not have enough energy. The simple fix is to ask more of the spacecraft. However, the spacecraft may not have enough energy either. The simple fix is, again, to ask more of the next element. Ask more of the lunar lander. And so on. With complex roots and causes, this poor habit explains parts of NASA's current lunar plans. The Gateway at the Moon results from limitations in the other parts and pieces. The debris from the lunar program's destruction fifteen years ago is still around today.

Another after-glow from NASA's earlier lunar program is the current Artemis program realizing they can't go it alone. There are elements of brinksmanship here again, of course, initially a European service module on Orion, or recently, the Japanese joining to provide a lunar rover. Elegantly, the Artemis Accords are a geo-political form of daring anyone to give too firm a critique. Words matter, and there is power in words and intent. As with the International Space Station, long ago, it faced a problem of stakeholder support, and the solution was finding more stakeholders. Too big to cancel mutates into too many stakeholders to stop. Yet a well-intentioned version of this effect offers hope. As General Eisenhower advised, "Whenever I run into a problem I can't solve, I always make it bigger."

"Whenever I run into a problem I can't solve, I always make it bigger. I can never solve it by trying to make it smaller, but if I make it big enough, I can begin to see the outlines of a solution."
-President Dwight D. Eisenhower

NASA has begun to make the problem of expanding our human presence to the stars bigger. The Artemis cis-lunar universe and an expanding list of Moon to Mars entries is a start. It's tempting to leave LEO behind, to the private sector and to private stations. Or a station. A mere couple hundred hours a year for NASA astronauts, on and off. Yet this assumes leaving LEO behind is an option. More likely, "you can't get there from here" will prove true as every step from Earth to orbit to the Moon and beyond accumulates the effort of previous steps. It's how gravity works and dollars too. As we put down the remote, we can say yes, *that's* how that works. LEO to Mars+ sees the step at low Earth orbit is critical and can't be skipped. "Moon to Mars" made the problem bigger, just not big enough.

* * *

Part 3

PART 3 SPACE TECHNOLOGY, R&D, AND OUR FUTURE EXPLORING SPACE

Employees in many industries could find themselves in an operational environment, "ops," or "production," it might be called, or many other ways of saying what's happening is a given - for now. The work is already laid out, and you must ensure it keeps humming. The space business has plenty of this, but it also has its earlier phase – creating change, innovating, and testing new ideas. However, Space technology faces unique challenges, unlike the pace of change in other tech sectors. If you want your new gizmo to fly one day, astronauts or not aboard, you will hear about the valley of death, how it weighs too much or lacks triple redundancy and a fail-safe mode. This will not be easy. After overcoming these problems, beware, as more barriers fall from the rafters. A new material? Refueling in space? Not invented here is a thing. Live long and prosper, and the bright side is the critics come around eventually to tell everyone they thought it was a good idea all along.

TECHNOLOGY STAGNATION AND NASA – PROBLEM AND OPPORTUNITY

My job with NASA always meant looking ahead. Today I can't help but look back. I am now retired, which I find an odd mix of calm, caffeinated, and a sense "I've seen things you people wouldn't believe." I arrived at Kennedy Space Center in 1988, a wonderful world of huge machines, buildings, and Space Shuttles –machines ahead of their times. And so many people. All abuzz to send our explorers to the stars. Everyone who contributed went along as well, at least in spirit. Soon after, walls fell I thought would stand forever. I could not wait to see what more was around the corner.

Thirty-three years later, we're sending astronauts to space again from KSC, after a 9-year stall, except fewer a year, in capsules, not Shuttles. We have the International Space Station, a wonder of technology and possibility. Our (so far) permanent presence in Earth orbit has people racking up more time learning to live and work in space than Shuttle's ever could. Yet plans for going further than low Earth orbit, people on the Moon or Mars with many people to follow, seem no closer now than when I first arrived at Kennedy.

Stagnation comes to mind, and it turns out this is not a problem unique to NASA.

Technology stagnation is a big red button we ignore on the console because we assume progress is inevitable. Tyler Cowen wonderfully captured this problem in "The Great Stagnation". A lack of innovation starves the supply side of an economy and leaves us endlessly only slightly improving what came before. Here there is no new and improved, just old and repackaged (something I'll try to avoid for myself). Cowen reminds us how real economic growth comes from real innovation. We should be amazed at the changes in daily life over time, not from fiddling with GDP numbers.

Who are we? Grandma, wide-eyed in 1975, describing amazing changes you people wouldn't believe, from the arrival of electricity, cars, and all those appliances, to advances in medicine, education, and a man on the Moon? Or will we try the wide-eyed look one day about media on demand? Apologists that technology stagnation is not a thing narrow in - the industrial revolution, then a transportation revolution (Grandma), then a communications revolution (us recently). More will follow. Patience. This is hardly filling, "sit and wait your next turn." Even the view there is no technology stagnation admits to pauses.

Jump ahead – what are the amazing changes in 2050 where life in 1975 is the old times? We're on the edge of great change. *I believe* space exploration and development can be a big part of looking back amazed in 2050. Maybe the hassle for the day is finding a room at Lianmeng station for the layover in route to USiCorp, one of many asteroid facilities that are no longer news.

Yet to make the story in 2050 about amazing change of the good kind means investments in space about growth through innovation. This means difficult shifts away from providers and stakeholders in economic environments that lack competition. Any space exploration plans, public or private, mustn't over-focus on boots-on-the-Moon or any single event with no plan for more. Getting there, staying there, and growing there will need competition to create innovations yet imagined, for dropping costs and prices, and for creating new customers and markets. This is all well beyond NASA's needs.

Exploring other worlds when "Earth is zoned residential and light industry" will require affordable, routine access to, through and from space for people, cargo, and ships. This is not new or impossible – we've been here in previous transformations. These transformations always mean changing who, how, and why. Start down this path, and it's not hard to see a world in 2050 looking back amazed at all the wonderful changes in our day-to-day life.

So, retired now after 32+ years with NASA, I have so many thoughts still to share from amazing times. Yet I also see many places we have been stuck, where innovation has stalled. As I said, my job with NASA always meant looking ahead, and I hope to keep doing that here in this blog. What role can NASA, the private sector, and space exploration play to ensure everyone looks back years from now amazed at the positive changes in their everyday lives? Let's explore.

OF EXTERNAL TANKS AND STARSHIPS

Iconic orange Space Shuttle External Tanks and shiny SpaceX Starships are uncannily close in scale. I was fortunate to be on the team in the 1990s that checked out and prepared the External Tanks and then on the team that filled and launched them. I could not have guessed that 23 years into my career at NASA, I would also be on a team studying refilling propellant tanks in space (my contribution being to connect the new technology with NASA budgets). Refueling in space is now the challenge that awaits the SpaceX Starship and NASA.

August 8, 2005, at the Space Shuttle Atlantis in the VAB. This stack was eventually de-stacked as the flight order after the loss of Columbia changed.

I often crawled around inside the Space Shuttle's External Tanks (and sometimes orbiters, and once inside a solid rocket motor – filled with propellant). I took the picture below inside a tank, in the "intertank," the space between the forward liquid oxygen tank and the aft liquid hydrogen tank. I recall the particular picture well for the care I had to exercise. This was not because I was inside a multi-million-dollar piece of flight hardware, and not watching my step could easily cause some damage. Not at all. Being inside flight hardware was by then a familiar and always conscious affair. Instead, my trepidation was because of the particular camera I held that day – one of the first-ever digital cameras, and NASA being NASA, we spared no expense.

In look and feel, the camera was similar to an analog film DSLR with a long lens – except this was 1991 and this was digital, and such things then were from the future. It was about $100,000 worth of technology from the future. Except as is visible from the photo, it has about the quality anyone gets today on a cell phone. That is from a cell phone that comes free with the monthly plan. The low-end cell phone. With the least expensive plan.

Flash forward and it's now nearly the tenth anniversary of the NASA Propellant Depot Study. The SpaceX Starship and refueling in space are poised to help lead the way to deep space exploration. Perhaps soon, we will see more of what is inside the Starships, apart from figures of the basic layout of liquid oxygen or liquid methane tanks and some feedlines. Perhaps we will see the diffusers, slosh baffles, crawl spaces, vent valves, and more. Perhaps we will also see that like that six-figure digital camera, things have improved.

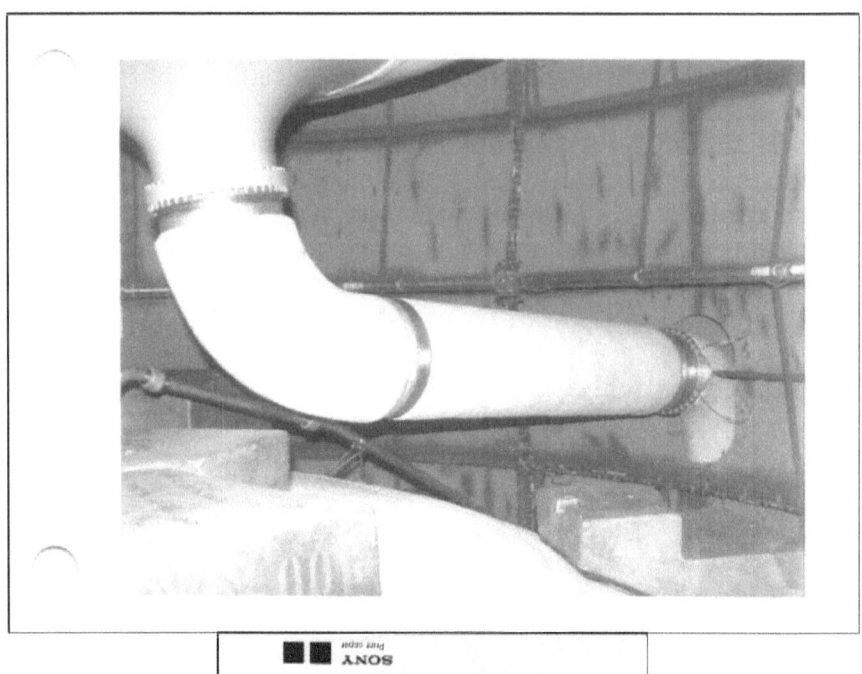

1991, External Tank "ET-43" interior, the "intertank" and photo print rear. The numbered blocks are light-weight but strong foam blocks to allow walking atop the dome of the tank. These are removed before flight.

AN ENGINEER'S JOURNEY IN NASA

External Tank

The External Tank (ET) serves a dual role: to provide the structural backbone of the Space Shuttle during launch operations and to contain and deliver liquid hydrogen (LH$_2$) and liquid oxygen (LO$_2$) propellants for the Orbiter's three main engines. The External Tank is 153.8 feet long and 27.6 feet in diameter. It weighs approximately 69,000 pounds empty and when loaded with propellants at launch weighs approximately 1,660,000 pounds.

The External Tank is separated from the Orbiter just short of orbital velocity at Orbiter's main engine cut-off (MECO). The External Tank pitches away from the Orbiter, tumbles, and breaks up upon reentry and falls within the designated ocean impact area. The External Tank is the only expendable element of the Space Shuttle vehicle. A further discussion of the External Tank is provided in subsequent sections of this handbook.

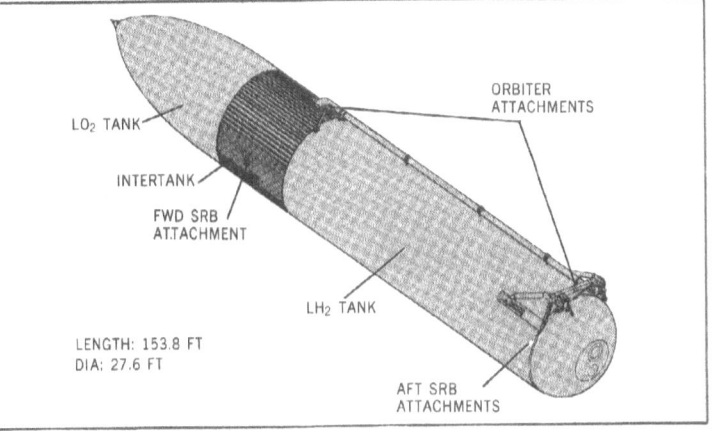

EXTERNAL TANK

II-5

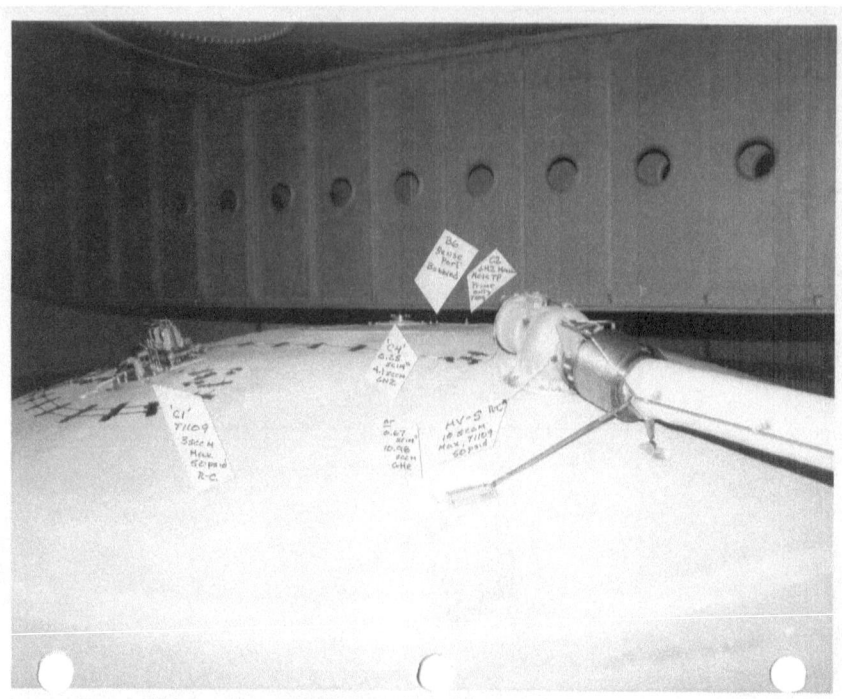

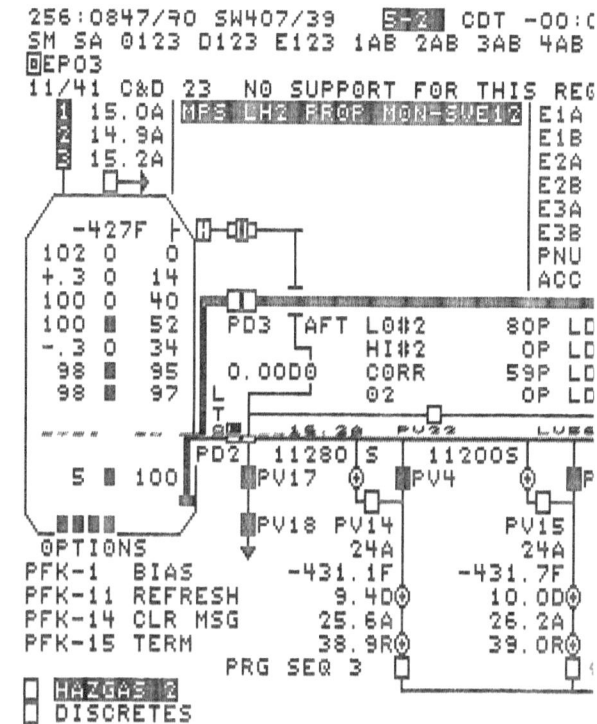

Prior - A page from the Space Shuttle External Tank handbook, showing the scale of the tank, a picture of the inside of an external tank (with annotations about leak checks), and a screen of the launch control computer while loading the external tank, from September 12, 1992, for STS-47 at Pad B, Endeavour's first launch attempt.

THE RISE, FALL AND RISE AGAIN OF REFUELING – IN SPACE

Range anxiety was invented by NASA. Well, perhaps not (or Velcro), but space exploration gives new meaning to an obsessive awareness of how much further you can go when there is no charger on every corner. Now imagine that feeling if you are in outer space or back on the ground watching your spacecraft, not just for power but fuel. In 2011, NASA looked seriously at refilling a rocket in Earth's orbit for the journey that was still to go. The team was asked to add meat to the bones of an idea that another NASA team had mentioned in passing the previous year as one option versus massive rockets.

Refilling stages in space makes too much sense. If low Earth orbit is halfway to anywhere in the solar system, and it was all hands on deck to figure out how to go beyond Earth orbit, then filling up before the other half of the trip had to be on the table. Yet the task did not end well for an assortment of reasons as events unfolded ten years ago. Not for the team, the technology, or its business approach. At least not just then.

If figuring out how to refill a spaceship in Earth orbit were a movie, after the opening where the characters are gathered up from their day jobs, the leader's pep talk ended with "...the secretary will disavow any knowledge of our actions". Even so, at the start, refueling in orbit was more than just promising. Technologically demanding yet manageable, with some vague benefits passes for promising. Here, though, the benefits were defined and significant. At the time, NASA was even funding real hardware with plans for a demonstration in space. As far as setting the stage, the NASA Constellation lunar return program had recently been canceled, unable to add up against likely budgets. With refueling in space, there was a chance to fit those lunar exploration plans back into foreseeable budgets. Refueling in space added up to exploration sooner, compared to other options that just didn't.

C'mon, it's completely impossible

Arthur C. Clarke is a favorite author of mine since I was gifted a book of his short stories by my 6th grade teacher Mr. Brown. Coincidentally, a story in the collection involves refueling in space (stealing the fuel). Clarke is quoted as saying revolutionary ideas go through three stages of reaction.

- It's completely impossible.
- It's possible but it's not worth doing.
- I said it was a good idea all along.

In 2011, the story about transferring propellant in space fell somewhere between Clarke's stages 1 and 2, between declarations of impossible or, at best and reluctantly, as possible but not worth doing. This was so even as the work did the math and answered the frequently asked questions, showing that transferring large amounts of cryogenic propellant in space was doable and well worth doing. It was a good idea! As innovative ideas often show convergence where more than one champion comes along at the same time, a team at United Launch Alliance led by Bernard Kutter reached the same conclusions. With flight-proven technology and those already in development, refueling in space was a game changer.

Technically, filling the tanks of a large rocket stage in space is nothing like filling them on Earth. Loading a Space Shuttle's external tank or any rocket tanks sitting on a launch pad has something a stage in orbit lacks – namely gravity. It's ironic how handy gravity is to load a rocket we use to overcome gravity. Yet the task of moving fluid between two sealed containers without gravity (not an easy trick) is well understood. It was never in doubt that it's possible to transfer super-cold cryogenic fluid from one sealed bottle to another sealed bottle ("ventless" or "no vent fill"). Think sealed heat pump, thermodynamics, and slowing down compared to the time it takes to load on Earth, and you are good to load and go.

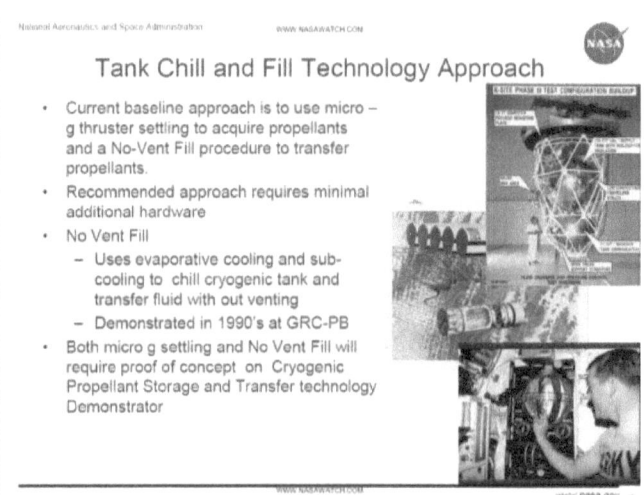

On Earth, the basic principle to fill a cryogenic tank is to open the receiving tank through most of the process. This open valve (called a "vent valve") is at the top of the tank. Liquid flows from a higher-pressure storage facility to the lower-pressure rocket tank. The loss of liquid out the top at the vent valve (boiled off as gas) is acceptable, as there is much more where that came from. In space, without gravity, up or down, top or bottom, other means are required. A sealed and "ventless" approach will ensure that almost all of the propellant pushed from a tanker remains inside the receiving tank. Credit: The 2011 NASA Propellant Depot Study.

Yet the demise of refueling in space after 2011 had nothing to do with technology as much as another game changer. New technology and its perceived readiness for prime time (or not) is often not a real issue against adoption, even if it's the first excuse offered. Everyone loves new toys. It's when technology is hooked to changes in ways of doing business that the barrier arises, now as if asking to change the laws of physics. An idea part and parcel to propellant transfer in space was going commercial for NASA's space transportation needs, not just propellant tankers.

Then along came another poor reason offered against refueling. While rockets are supposed to get cheaper as they get bigger, this is only true for apples to apples, say comparing two "cost-plus" programs (the old business model, NASA owned, no ongoing competition, NASA as the only customer). This notion a larger truck will haul cargo more cheaply than a smaller one – cost as measured by the pound – does not apply when comparing a large commercial rocket, as the propellant depot team was doing with Falcon Heavy, to an only marginally larger cost-plus rocket. Not even close. The relationship is then reversed – the only slightly smaller commercial and very competitive systems do better on all measures. Nonetheless, this ill-conceived notion *any* larger rocket must be better by the pound, among other seemingly technical criticisms, spelled the demise of refueling at the time.

Well, it's possible but it's not worth doing.

Ideas are bulletproof, especially excellent ideas. The work a decade ago heralded in the press as "gas stations in space" came and went, and the studies and the hardware scrapped from further consideration. Another team would come along in 2015 for a second bite at the apple. Some of the 2011 team (myself included) were at it again. Here, the gas station in orbit was ditched, and the space tankers would just refill the stage in orbit directly. Imagine a couple of gasoline tanker trucks filling up another truck. Rendezvous, mate, and fill-er-up. (In my DOD circles, I heard a sigh of relief that the depot, the "first big target in a war," was gone from this equation). Also, the refueling was no longer central in the 2015 work, necessary to the task, but with no need to overly defend it as feasible technology. Instead, the report went to the heart of the matter, new ways of doing business. This was, after all, the hard part for many in the target audience. The new ways of doing business could create a propellant economy in space, starting in Earth orbit. Yet, as with the earlier effort, this work came and went. Its value remains in the persistence of memory, spreading like so many thoughts that leave an impression.

You know, I said it was a good idea all along!

It's now 2021, and refueling is at the stage where it's common to hear, "I said it was a good idea all along." NASA is back in the refueling business. It may have had something to do with physics, about Earth orbit being the first most difficult climb out of a gravity well, and asking, "If only we had a full tank." More so, it may be how new (commercial) ways of doing business have taken ground across all the new NASA programs – from low Earth orbit to the Moon. It may also be ideas like refueling stages in Earth orbit have to pass through Clarke's stages of revolutionary ideas, fueling up along the way.

WHAT'S OLD IS NEW AGAIN – MORE ON REFUELING IN SPACE

On my shelves sits a childhood book "Planets and Spaceflight," published in 1957 by General Mills. The front cover is "Planets," and the rear is "Spaceflight," full of vivid descriptions and beautiful artwork of the many places to go and how we will get there. The publisher is best known for Cheerios, so I'm sure my parents picked up the book at the supermarket as part of a deal. Was this like those old encyclopedia sets, "A thru Amer," for 25 cents with $10 in groceries? I happened on the book just days ago, only to see it included refueling in space. This was that proverbial moment in an uncanny valley, having worked on in-space refueling in NASA a lifetime later. NASA teams (2011), later teams (2015), and others outside NASA (2008 and 2012) have reduced refueling in space to its numbers and graphs, minus the inspiring artwork and the musty but somehow comforting smell of an old book.

General Mills "Planets and Spaceflight", 1957, by Otto Binder, Illustrated by George Solonevich.

Today, NASA is investing heavily in refueling, an idea that was once verboten but is now in vogue again. This may be thanks to a massive Starship passing as just a lander. Starship "refilling," the wording better reflecting that both fuel and oxidizer are topped-off, plans to trade the propellant not in one ship for the payload not placed on another. The imagery of what's planned goes well alongside the artwork of days past, reminiscent of tank after tank all in a row. As bookends on a journey, refueling in space has come a long way. Yet, there is actual hardware that can tell a lot about in-space refueling. Between the art of Solonevich and the plans of SpaceX, there was a stage called the Saturn S-IVB.

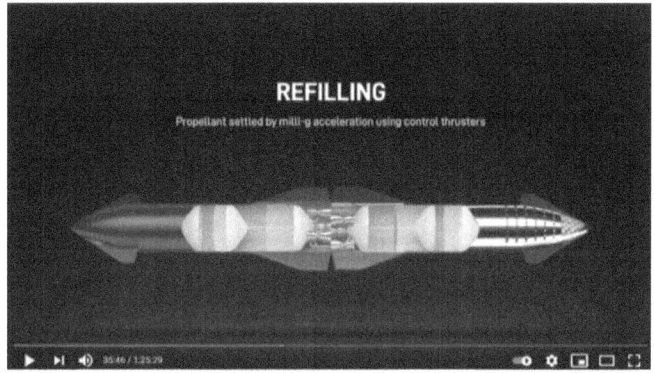

SpaceX Starship refilling. Credit: SpaceX.

Below, the Saturn S-IVB, actually a 3rd stage, not a 4th, is afloat against a cloudy sky with a lunar lander attached. I see this and hear "Mr. Sulu, assume standard orbit" in the background. At first glance, the picture would appear to be a stage awaiting the Apollo crew spacecraft, which is turning around to mate with the lunar lander and access port to access port. This is the case. It's not for a lunar mission but for the Apollo 9 test flight of everything, from stage to lander to spacecraft, with even some extra-vehicular activity. The scale is telling, a stage large enough to get landers and spacecraft (and adapters) along the way to the Moon had it fired up its engine. Notably, when the lunar missions did leave Earth's orbit, the stage was not entirely full. A lot of propellant had been burned to get it into its proper orbit. Imagine if such a stage had been refilled. Now imagine again how an empty or near-empty stage of nearly the same scale and a lander attached could be launched without a massive Saturn rocket.

The Saturn S-IVB stage and lander during the Apollo 9 test flight. NASA.

Starships may set the stage for refueling, but refueling can excel at much smaller scales. While we have a very commercial US launch capability today in the SpaceX Falcon 9, we also have a very competitive launch capability on the smaller end in Rocket Lab's Electron rocket. Others will follow. A single scale to meet all needs would also seem unlikely for deep space. Might refilling also find a helpful scale below the Starships? It turns out 100,000kg of propellant, the scale of the Saturn S-IVB, is a minimal scale for crewed lunar missions. Between the time of the childhood artwork and the recent plans for massive rockets, there was real hardware that points to possibilities.

(See the "Saturn V Launch Vehicle Flight Evaluation Report-AS-506 Apollo 11 Mission" at https://www.ibiblio.org/apollo/Documents/lvfea-AS506-Apollo11.pdf, pp. 108. The original estimated amount of propellant in the Saturn S-IVB at engine ignition (in space already) was 87,315kg of liquid oxygen and 19,780kg of liquid hydrogen. The amount remaining at ignition to leave Earth orbit hours later was 61,300kg and 14,395kg.)

To say refilling a stage in Earth orbit has many moving parts would be an understatement. All the parts begin their story by scaling a stage. It's no coincidence the other current development, the NASA Exploration Upper Stage, is close in scale to the Saturn S-IVB. Orbital mechanics and physics are a constant. If the smallish Apollo lunar lander was about 17,000kg fully loaded, and an empty stage to be refilled later is around 12,000kg, there is now an extremely affordable launch capability to place that tonnage in low Earth orbit. Better yet, in a growing economic ecosystem, there will be other capable launchers to follow. Say a Falcon Heavy can place 63,800kg into low Earth orbit, then add the caveats. If we wish to use it in fully reusable mode, it's more like a 34,000kg capability. If we want to place the ship to await tankers up higher, at 400km, and say at 51.6 degrees inclination (like the International Space Station), we must work within about a 30,000kg launch capability. That empty stage and a small lander add up – but with too little margin.

However, there are plenty more puts and takes. A stage doesn't have to be as large as the Saturn S-IVB. It's already been placed about where it should be. More so, expending a center Falcon booster and reusing only the side boosters, still a great cost advantage, quickly jumps total capability for that initial mass put at 400km/51.6° to around 51,300kg. Now that stage can get bigger again, and the lander and other elements like the spacecraft service module can scale up, too. Some initial propellant might be added, just for starters. Adapters, maneuvering, losses and other factors regain ample margin to spare.

From these masses and the physics, there come possibilities in the cost, safety, and reliability improvements from NASA's partnerships, buying launches, and propellant as competitive commercial services. This is so even when NASA is the only customer in the near term. It takes a while to get to Inspiration4 using commercial crew spacecraft developed for NASA, but well before and regardless, there is every indication NASA reaps enormous benefits. Time will tell how refilling will play a role in exploring our solar system – but it appears the stage is set so a ship in space filling up on propellant before its long journey is no longer just the stuff of childhood stories.

ONE WORD: PROPELLANT

Some graphs, like pictures, are also worth a thousand words. They do what a beautiful painting does while wandering in a museum, holding your stare like reading from a wall. For rockets and space travel, there is no shortage of figures and numbers and graphs, oh my. One especially telling figure came around in 2010 during a NASA look at refueling rockets in orbit. Suppose you are trying to galivant around the galaxy, but having only reached Earth orbit, you already have less than a full tank of gas. Bummer. The trip only started, and the needle already dropped.

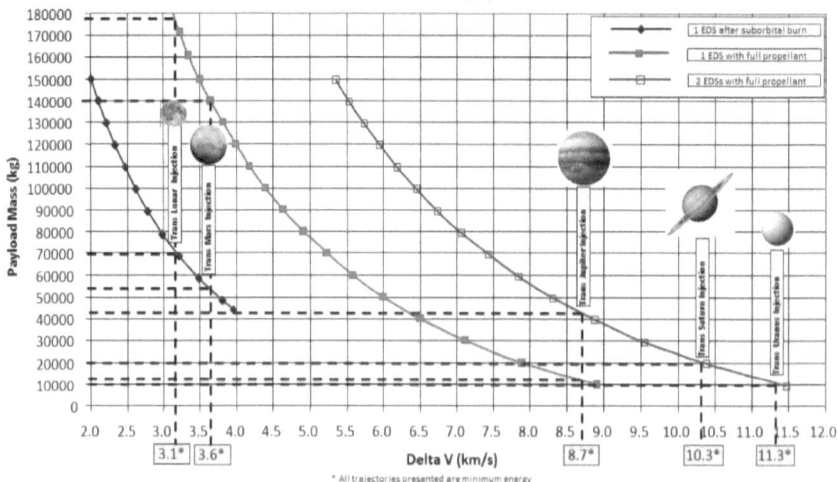

A propellant depot-based architecture, capabilities from refilling a stage. Wilhite, Arney, Jones, Chai, 2012, Georgia Institute of Technology.

Of course, there are options, like making your rocket bigger, so what initial propellant for your trip is placed in orbit is that much more. NASA already did this in the Apollo Moon program decades ago, and it will do so again with its new Space Launch System. You can also break up what you need for the trip, with parts of it launched at different times. You meet up with the truck in your car. This too is part of NASA's current plans. Yet, there is another alternative. Fill up your tanks before you leave Earth orbit, tanks you might even send up mostly empty, so more easily, on smaller rockets.

"It's all about delta-v" is a phrase that will always hold true for space travel, any year, any time. Even as technology advances and we come up with new buzz words every year, saying it's all about 3D printing, or AI, mining, pharma, or SPACs, saying it's all about delta-v will never get old. How much can you change your velocity in some direction? Your delta-v capability is a number that says how much the gas in your tank can do for you. Where can you go? And always a good idea with a crew, do you have enough to come back?

Our NASA community famously enjoys presenting the kind of chart with a few more moving parts than just *if this, then that*. There is always something else, often buried in a lengthy narrative. But my translation of the beautiful chart above, in simple terms, is - *if you have seen that series called "The Expanse," we are getting awful close.* (Excepting the nuclear fusion technology on their most advanced ships.) If we can just get more propellant where we need it, when we need it, cheaply, space opens up in ways that put our outer planets within reach. This is not for the usual robotic probes but much larger and more capable machines freed up from the limitations of watching their weight down to every last ounce. We even open up space beyond Mars to crew. Asteroids anyone?

Consider a spaceship with a propellant tank of a size to fit on a Falcon Heavy rocket. Or perhaps we break this up into two smaller tanks. We launch the smaller tanks on any assortment of affordable launchers to come from United Launch Alliance, Blue Origin, or other new launch providers. Or perhaps we join up small and large tanks in orbit. Mating up is usually a part of NASA's plans for deep space exploration, some assembly required. Imagine if putting a pound of anything in orbit continues to drop, knowing the one thing most needed in orbit is propellant.

To get to Mars (or anywhere) from Earth, you need more propellant than spaceship. Think of it as your car with a fuel tank bigger than the cabin you sit in, and every other part of the car too. But even as transporting a pound of anything to orbit drops in cost, a pound of propellant and a pound of space systems hardware are two very different things. The former is simple, the latter complex. Different, though, does not mean propellant and hardware are independent. On the contrary, how propellant and space hardware are linked opens possibilities.

Now imagine a long deep-space voyage, not because your ride is slow, not because you can't go faster if you wanted to, but because your purpose in being out there is not to just hit and run. The bane of such dreams is dual-fold - space radiation and zero-gravity, each bad for your health. The one being present wants to hit you with particles, or all manner of radiation, many times the doses anyone should safely get in their lifetime. The one being missing, a lack of gravity encourages your body to go haywire starting at the cellular level, and moving on to just about everything else – bones, eyes, liver, and muscle. Yet you can now get anything to Earth orbit much cheaper, and so you can also get much more propellant, to push that mass out of orbit. So, start your trip right, with a full tank.

Artificial gravity schemes, and ever more shielding across more of your spaceship, are suddenly *not* something to discard as too heavy to get to orbit, *or to take along.*

Shielding for crew safety means mass, with passive additional materials, or active schemes deflecting harmful particles. That mass must be propelled as we explore – which brings us back to propellant. Power comes along as well. Having more propellant, you no longer have to be miserly about power. This is propellant translating into electricity because you could add the mass of more solar panels or nuclear generators. Gravity schemes that rotate a spaceship and living quarters also mean more mass. Whereas today such concepts try to make a go of it on a diet, assuming current limitations on launch and the propellant available, tomorrow they may be liberated of these chains. As an added bonus, along comes reusability, because why toss such expensive hardware as if it's a one and done, when you can just refuel and use it again.

Graphs about your delta-v to get to Jupiter are definitely complex, but the connections are not. More and cheaper propellant in orbit, and refueling before leaving, means being freed of the limits of lightweight thinking. That one word, propellant, is really so many more words - more spaceship, more power, more crew, and longer, yet safer voyages out there. The reusable SpaceX Starship already sees this, with plans to refuel in Earth orbit before a trip to the Moon. The matter ahead is who else sees that one-word *propellant*, and all the other words that follow.

STARSHIPS MEAN REFILLING YOUR TANK, AND SO MUCH MORE

NASA press releases often come and go where the world is left to ponder a message one step removed from chicken bones strewn on the floor-mat. If it's not the acronyms, it's the lingo or the leaning to put out only the facts, not what they mean. But if NASA ever buried the lede, it was on April 16, 2021, announcing the selection of SpaceX to build its lunar lander. A year and a half later, the choice of such a massive ship for astronauts to land on the Moon still overloads the conversation. You can't stop hearing Jan saying Starship, Starship, Starship! So much more has gone unsaid, buried in a few threads, meetings, and obtuse NASA documents. With the Starship lunar lander, NASA will also get in-space refueling, so propellant tankers. Then there is the magic phrase "and wait for it, if you order now," where we also get a refueling station in space, and, wait for it again, possibly space commerce that will make the rest pale by comparison.

SpaceX lunar lander selected by NASA. Credit SpaceX.

A large lander that hides the real stories behind it is not new. Long ago, NASA had another lander on the drawing board, causing a stir for not quite what it seemed. This semblance is so, even as the lander then and the lander (err, Starship) today could not be more different. The common thread is the world these spaceships live in, with laws of physics and limits to NASA's budgets.

In NASA's last plan for a return to the Moon, the Constellation program drew up a lander that, like other parts of the project, screamed "Apollo on steroids." The Altair lander was born from ambitious goals, and its size said as much. After all, the lunar return plan was about a lunar outpost to stay, more than a series of quick visits. Altair would stretch NASA's plans to their limit, and past the breaking point, at two or three times as large as the little Apollo lander that first took astronauts to the Moon. Now imagine a meeting where a lowly engineer asks in a Tom Hanks, "I don't get it" kind of moment, "where is our lunar outpost?"

Some blank stares later, and in more meetings to follow, no one wants to put the lunar outpost on the schedule vs. funding. Any outpost. I'm told, "not yet." The budget over time is up on the screen, a basic profile, but still, advertised as the plan. There is a star icon far to the right with an unremarkable note like "crew landing." We show launchers, spacecraft, a lander, and other big-ticket items like the International Space Station. Yes, the ISS was in the last Moon plan. "But where is the permanent outpost, delivered by that lander?" I asked.

Obvious as it seems, I thought I was missing something, except I wasn't. Months later, probably not from persistence on my part, an outpost wedge showing funding years before the advertised lander delivered it was (finally) added. This edit came in time for murmurs of a committee, then a program cancelation. Such odd behavior was likely a symptom of why this last plan's demise was inevitable. This obvious addition should not have been that difficult.

Everyone knew the real reason a lander delivering an outpost seemed an aside in the last lunar landing program. There was no funding and, worse, no plan for funding, not for a lander, and even less for the critical goal of the whole project, an outpost. Being broke is not a fatal flaw, a lander being many years away and an outpost even further. (We would have returned to the Moon a couple of years ago.) The fatal flaw was the lack of interest in building a story that could credibly show how funding might show *and add up* in the big picture. I also knew this but believed putting all steps, even ones no one had plans for, on a schedule to stay on the Moon would begin the conversation about what we must do differently.

Also, and unbelievably, a thought stayed on the shelf way past its due date where funds for the lunar lander (and more) would come from de-orbiting the ISS five years after we completed its construction. This notion was easily removed from early schedules but not from some hearts and minds. None of this about planning not adding up, and more poor assumptions, made the headlines alongside pictures of the huge lander, but the lander is what bubbled them up. The lander was big, and so was everything else, and with time we would return to the Moon to stay. That buzz buried the real ledes then, for a time, until they all unburied themselves at the cemetery and wandered everyone's way.

* * *

through at least 2012. Completion is now scheduled for 2010 or perhaps early 2011. As recently as 2003, NASA briefing charts showed operations possibly continuing through 2022. Under the Vision, U.S. utilization is scheduled to end after 2015, but widespread efforts to extend that date are ongoing.

Congressional Research Service report from 2010 where the ISS had a "currently planned termination at the end of 2015"

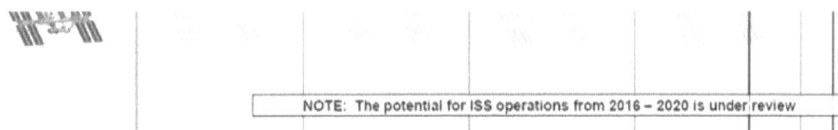

NASA's schedule from 2007 carried on an implied and flawed assumption of an early ISS demise. "NOTE: The potential for ISS operations from 2016 - 2020 is under review."

Today's NASA lander will be a Starship, but the fanfare is not hiding bad news. Quite the contrary, this is a Starship for only a few billion dollars out of NASA's coffers. That's the kind of money NASA civil servants find between the sofa cushions in the Director's office. Well, maybe the sofas in DC, among the Associate Administrators. For comparison, the commercial crew Dragon spacecraft cost NASA about as much to develop, $2.5 billion (in 2022 dollars). For starters, a headline that did not bury the lede a year and a half ago would have been "NASA buys lunar lander for *only* $3 billion." Yet even this misses the mark.

With the Starship lunar lander, you get in-space refueling. The idea is simple. Gravity gives us Earthlings only a hand full of options for exploring our neighborhood. One, you arrive in Earth orbit running on fumes. That is unless you have a giant rocket putting in orbit a rather large throw-away stage with enough fuel to push a small ship to the Moon (that must also have its own fuel to come back.) Two, you can put parts and pieces in orbit and meet up, or hook them up, now able to go places. (Today's NASA plan is a variation on this.) Or third, you can fill-er up.

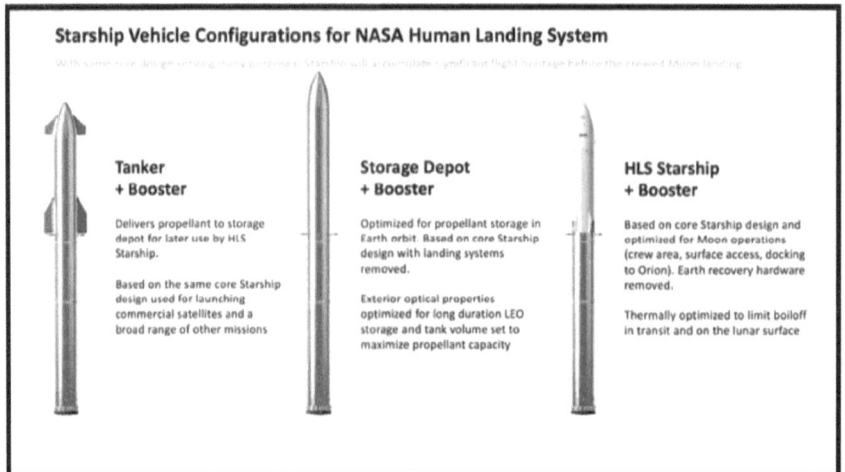

Fig. 2. The three Starship variants that will be utilized for Artemis III are Tanker Starships to transport propellent, a Storage Depot to store the propellent in Earth orbit, and the HLS Starship that will travel to the Moon.

From September 2022, more information on the SpaceX Starship as NASA's lunar lander, which means Starship tankers and a Starship propellant depot. NASA, "NASA's Initial Artemis Human Landing System," Dr. Lisa Watson-Morgan, Mrs. Lakiesha Hawkins, Mr. John Crisler, Mr. Larry Gagliano, Mr. Rene Ortega, Dr. Thomas K. Percy, Ms. Tara Polsgrove, Mr. Joe Vermette.

> While I find the positive aspects of SpaceX's technical approach to be notably thoughtful and meritorious, these aspects are, however, tempered by its complexity and relatively high-risk nature. Of concern here is the SEP's assignment of a significant weakness within SpaceX's proposal under Technical Area of Focus 5, Launch and Mission Operations, due to SpaceX's complicated concept of operations. SpaceX's mission depends upon an operations approach of unprecedented pace, scale, and synchronized movement of the vehicles in its architecture. This includes a significant number of vehicle launches in rapid succession, the refurbishment and reuse of those vehicles, and numerous in-space cryogenic propellant transfer events. I acknowledge the immense complexity and heightened risk associated with the very high number of events necessary to execute the front end of SpaceX's mission, and this complexity largely translates into increased risk of operational schedule delays. However, these concerns are tempered because they entail operational risks in Earth orbit that can be overcome more easily than in lunar orbit, where an unexpected event would create a much higher risk to loss of mission.

2021 NASA selection statement for the SpaceX Starship lunar lander.

NASA is no stranger to the idea of refueling in space (nor am I). After the last Moon plan's demise, NASA chartered a team to see what might be salvaged and add up inside realistic NASA budget prospects. At the time, we began with a refueling station in space, a "depot" filled by "tanker" spacecraft. The ship that would leave Earth orbit followed, filling up at the gas station before heading out to the Moon or further. The elegance was in using existing or up-and-coming launchers, creating a market for a commodity provided to NASA on a commercial basis. This idea was not about delivering people or complex one-of-a-kind satellites but addressing the simplest, largest mass necessary to leave Earth orbit - propellant.

The scale we looked at ranged from 100 tons (metric) to over 200 tons of propellant (fuel and oxidizer, though often said as "refueling.") In reverse, and to simplify, we later explored refilling a stage directly from the tankers, skipping the depot middle-man. Imagine fueling your car from a small tanker truck, minus the gas station.

For comparison, NASA's new Starship lander will be in a field of its own – clocking in at a massive 1,200 tons of propellant, and perhaps transferring 100 tons or more of propellant per trip. We can imagine a Starship depot will similarly hold about 1,200 tons of propellant to fill a Starship to leave for the Moon. That's over *ten times* the size of the smaller refueling station in space NASA looked at in 2011. In reverse from our evolution a decade ago, SpaceX began with the notion of refueling direct from tankers to stage, but they now see the need for a middle-man depot to service such a large-scale Starship. So launch your lander Starship *after* it can fill up *all at once* and don't keep this pricier and crew-capable Starship variant waiting for tanker after tanker.

* * *

Starship tankers, Starship depots, and Starship lunar landers – *but wait, there's more.* Would anyone else like to fill-er-up? If you have 1,200 tons of propellant in low Earth orbit, might you decide to sell some to others? After all, NASA's lunar missions may not buy up all the fuel and oxidizer available. When NASA first launched astronauts directly to the Moon, the stage that entered Earth orbit with the small Apollo spacecraft and lunar lander needed about 75 tons of propellant. Some years from now, we may have at the ready 16X this amount, just begging to fill up other large stages, all put in orbit more efficiently, being *near empty when launched*.

So as not to bury the lede, imagine the Starship-lander announcement had been "NASA buys lunar lander for *only* $3 billion, and gets a place to refill rockets in space too!" The news clip would add a little history going back to 2011 when refueling in space was verboten, then jump to six months earlier, as NASA partnered with companies to work on in-space propellant transfer. All this would be a lot to fit in a headline, not as short and likely to draw attention as "man bites dog." A Starship, a return to the Moon, changes in policy and politics, the limits of physics and budgets and depots oh my. Last time, a lander showed the breaks in NASA's plans to land on the Moon. Today we have a lunar lander that breaks all the rules. We can look forward to the headlines.

REUSING, REFUELING, PARTNERING - AND GOING NUCLEAR

Advocacy for innovation is always challenging, with much written about difficulties like the valley of death. There is one barrier that does not get much attention, though. We forget the future is always outnumbered in the here and now.

This seems obvious but does not get the press it deserves. Advocates for what's new are, at best, stand-ins for a future organization, vast but vague. These other people may (or may not) exist one day, manufacturing and operating the new product. The initial effort, data, and its advocates are the ghosts of Christmas future, showing what might be to an incredulous Scrooge. Except here, the spirits reveal a bright future and ask you to help make it so.

Franklin C. Spinney was among the first to dare say the obvious implications of a fact of life – an existing, well-funded project is at an advantage, using its funding to get more funding and to ensure it sticks around. Then, as if that's not enough, they insert themselves into the meetings about what's next. This becomes your audience for that incredible but new technology. An already established project buys a seat at the table to decide the future. Worse, and too often, they own the table.

This week NASA and DARPA announced they would work together on a new nuclear thermal engine to be tested in orbit. Why nuclear? Well, a more powerful engine will reduce trip times to Mars, or anywhere for that matter. Faster trips mean less time for the crew in zero gravity with its well-known adverse health effects.

Artist concept of Demonstration for Rocket to Agile Cislunar Operations (DRACO) spacecraft, which will demonstrate a nuclear thermal rocket engine. Nuclear thermal propulsion technology could be used for future NASA crewed missions to Mars. Credit: DARPA.

Faster, shorter trips also mean less accumulated radiation exposure for the crew. Radiation adds to health risks as if the effects from the lack of gravity were not harmful enough. On the International Space Station, the Earth's magnetic field provides protection from radiation (though not altogether.) So, as a rule, if you leave Earth's orbit and its magnetic field, *make it quick*.

None of this is new, as speedier nuclear propulsion was NASA's plan as recently as 2009. We will see if the NASA/DARPA announcement takes on substance this time or remains aspirational.

> 4.2.3 Earth Departure Stage as the Trans-Mars Insertion Stage: Commercial Tanker
>
> This option is comparable to the previously mentioned propellant tanker option, but it reduces the number of required Ares V launches by utilizing commercial tankers. As the Ares V propellant tanker option evolved, it was noticed that most of the required launches are dedicated propellant delivery launches for the TMI stages (five of the nine launches). As a potentially enhancing capability, it became evident that if a commercial market for LOX/LH₂ propellants indeed exists 20-25 yr in the future, the dedicated Ares V launches to deliver this propellant could be replaced by this commercial market. Therefore, the absolute minimum number of required launches for Ares V was found, which was four. This consists of the very large masses and the platform for the commercial tanker services to deliver the propellants to (i.e., empty or nearly empty EDSs). In addition, there will be numerous, but not yet determined, commercial launches necessary to provide the additional propellant to the TMI stages for the MTVs.

NASA, "Ares V Utilization in Support of a Human Mission to Mars," 2010. The Ares V morphs to become the renamed NASA Space Launch System.

Yet a lot has changed since 2009. Nuclear propulsion will now compete with other means to the same ends (and none of these other paths has the words "nuclear" or "uranium" drawing regulatory attention.) At its simplest, a lower price for NASA (or anyone) to get more mass to orbit could reduce crew health hazards in ways other than nuclear engines and their faster trip times. More (cheaper) mass to orbit could mean more propellant, which could also mean faster trips. This is especially so with in-space refueling for your spaceships. This is in development by many companies, most notably for SpaceX Starships, one of which will serve as a NASA lunar lander. More, cheaper mass could also mean adding more radiation shielding. Lastly, a holy grail, the ease of putting mass into orbit opens up the possibility for rotating structures creating artificial gravity.

But let's go for broke. More propellant, more shielding, and maybe even gravity, *and* all of it going faster with (fission) nuclear engines – what's not to like? (Did someone say fusion? Even better.)

※ ※ ※

And yet, as the NASA and DARPA "nuclear" announcement trended, I could not help but see a similarity to other well-justified paths that still faced enormous difficulties. Just because it all sounds so good does not mean everyone comes running to help. Consider reusability and launch. If ever there was a conflict of interest, it was a room filled with people debating more affordable, reusable launch systems, launching much more often, with most of the people invested in expensive, expendable hardware. It should surprise no one that the decades-long decision-making around what would come after the semi-reusable Space Shuttle went as it went.

First, we saw (airbreathing) spaceplanes ahead, then single large reusable rockets, then two (even three) smaller reusable stages, and finally only a tiny reusable spacecraft – the Orbital Space Plane, atop an expendable rocket. The last death throe modified the length to diameter, or L/D, and turned the OSP into an expendable capsule, which became Orion. Reusability was crushed in every cycle, literally.

A casual observer would say that reusable Space Shuttles, the orbiters, and the workforces at Johnson and Kennedy were well invested in reusability. These were no small armies in the field of battle. That was the unstated assumption ending the talk about balance among groups advocating one direction or another. Missed until it was too late was realizing no one made new Shuttle orbiters. Lacking manufacturing, operators were quickly seen operating anything if the thought was keeping Johnson or Kennedy busy. Full reuse in a new system would eliminate most parts of the older Shuttle program, but a system throwing everything away (but based on Shuttle) would not. So why not keep everyone around – with a new, expendable rocket.

Similarly, refueling ships in space was an idea whose time had come. (Refueling is a close cousin to reusable launchers, opening the way to reusing hardware already in space.) Curiously, refueling appears in the same report as the nuclear propulsion option for Mars. But as stakeholders go, we learned in 2010 why they are called "stake" holders. Chartered by NASA, our team accepted the mission knowing if we failed, the secretary would disavow any knowledge of our actions. Fail we did. And disavowed we were. Again, selling a great new technology to decision makers that are also stakeholders seeing themselves losing out from the change gives new meaning to "a tough sell."

* * *

Yet time has a way of proving that physics is what it is, and dollars must add up too. By the oddest turn of events, NASA invested in a small company called SpaceX. They turned an initially expendable first stage into a reusable booster via landing legs and figured out how to reuse payload shrouds too. With their Starship, they hope to make a fully reusable launcher, booster, orbital spaceship, and all.

As you can't keep a good idea down, refueling has also made a comeback. SpaceX and other companies like Eta Space, Lockheed, and United Launch Alliance are working with NASA to develop in-space refueling technology. This is the technology the mere mention of which became verboten in NASA over a decade ago. Limits on the size of any one launch mean the amount of propellant and spacecraft that might leave Earth orbit is also limited. To get around this, assembly in space follows, inevitably, as does topping off your tanks. Eventually, reuse again follows, refilling the same tanks on the same ships in orbit, over and over.

Yet the most remarkable breakthrough in NASA (or even SpaceX) in recent years may not be a technology as much as a relationship. The moment NASA signed its first partnerships to get cargo to the ISS should have been noted, like the launch of a new rocket or fresh pictures from the Webb telescope. Going commercial should be up there with going nuclear. Physics helps keep great ideas alive, but incentives make them grow. The right audience, decision-makers, and stakeholders are like rich soil for planting seeds. Eliminate the conflicts of interest, align incentives properly, assure a growing space sector is everyone's goal, and the garden will prosper.

The writer Upton Sinclair said – "it is difficult to get a man to understand something, when his salary depends on his not understanding it." For going nuclear when going to Mars, the path ahead may be just as convoluted as it has been for reusability, in-space refueling, or NASA going commercial. And the story on these is still being written. What I can share is never to be surprised by how things work out, if the stars align, and incentives.

IT'S A SYSTEM

Rocket launches, a possible boil water notice here in Orlando, and hospitals caring for patients with COVID are all connected. Now, it's about liquid oxygen, but finding more connections would not be surprising, like in any system. Oddly and often in projects, "it's a system" was an observation that arrived at the party early only to find no one ready to receive it. If later, we were told everyone was now entertaining others. Coincidentally, I worked for some years in both liquid oxygen facilities for the Space Shuttle and a systems engineering office. Now, any news day is like the cat scene in The Matrix, that feeling of déjà vu, then brace for incoming.

The Space Shuttle Launch Pad A liquid oxygen storage sphere viewed from about mid-way to the launch pad. I took this picture during a cross country line walk down in 1996.

It's a hot and humid Florida day, mid-morning, and worse, everyone is in heavy denim fire-retardant coveralls. Five liquid oxygen tanker trucks park neatly in reverse, one for each valve on a fill manifold to the Space Shuttle's liquid oxygen (or "LOX") storage facility. The drivers pressurize their tankers, opening and closing valves at the truck's rear housing. They connect hoses, and a welcome snowy frost covers these as liquid flows in them at *minus* 297 degrees Fahrenheit. The sea breeze cools, a welcome if nearly imperceptible degree, from passing over the frosty lines and if you stand in just the right spot. Once the offload is complete, the drivers disconnect the hoses, and residual droplets of LOX fall to a fizzle on the cement. We will do this many times throughout the day, to fill up the storage tank and stand ready for our next launch.

Thinking in terms of parts was fine, for a time.

After spending some years working on the Shuttle's bright orange External Tank, I saw more connections. On launch day, we filled this tank with LOX (and liquid hydrogen). I started to appreciate by then - it's all connected, all parts of a single thing.

Eventually, I found myself in the Kennedy Space Center's "systems engineering" office. No one had a good idea at the time what the job title meant. Our discussions meandered like

Russian babushka dolls, one inside another. The avionics is a system, right? Or is the system avionics plus the body flap? Eventually, an understanding came around – like the boy who cried wolf – and I reserved saying, "We need to think about this as a single system" for problems we could no longer afford to think about in parts.

The fuselage of a reusable launch vehicle was among the first wolves we spotted. Before, it was possible to think aluminum airframe here, thermal tile over there, and lots of coordination and goo in between. However, whereas the Shuttle had an external tank, a future reusable launch vehicle, like the SpaceX Starship today, has internal tanks. Inevitably, the new arrangement of materials was now a system. You could look at it as parts here and over there but at your own peril. First is the structure (composite, aluminum, or stainless steel, as with Starship). Then, there is insulation for the super-cold liquid oxygen or methane tanks. Lastly, there is thermal protection for a fiery return from orbit, perhaps like the iconic Shuttle tiles. (And some thin layers I'm skipping over.) It is not as simple as it sounds – ponder how to stick a shower tile *onto the wall insulation, not the cement board.*

The "sandwich" (as I came to call it) of materials that emerged was a system. It was no longer possible to think of this bunch of items and people as pieces of a puzzle that come together at the end. The parts needed to come together at the start. Public information does not do credit to the technology and its difficulties.

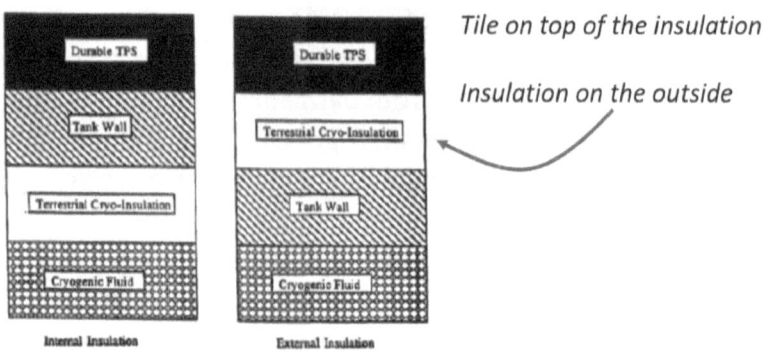

National Academies Press, Reusable Launch Vehicle: Technology Development and Test Program, 1995.

Yet, thinking of a sandwich of materials, steel, quilting, maybe foam, tiles, primers and goo and more, as a system is simple – compared to a ship you want to fly to orbit *like an airplane*. An airbreathing spaceplane with a hypersonic engine was another challenge for NASA and DOD projects. Again, it slowly became the consensus that it was all better viewed as a whole from the start. Breaking a problem into manageable steps or pieces is usually good until it's not. That was the case here – unlike with a traditional rocket engine, we had connections that could not be conveniently severed to parse work. Typically, there is an engine and the ship structure, all this meaning people, with other technology and people in between. Unfortunately, at best, "integrating," as we said, was a fancy phrase about making sure all these people stayed out of each other's hair.

None of that will do for a true spaceplane. One day, if you were flying to orbit, like Heywood Floyd in 2001: A Space Odyssey, it wouldn't have happened because someone broke up this new technology into smaller, more manageable pieces. This, too, will be a system, the parts all connected to where you can no longer easily say where one ends and the other begins – structure, engines, and keeping it all from turning into a molten mess.

These spaceplanes will use less liquid oxygen because, like

airplanes, they will use oxygen from the air they fly through. But we must hope we don't have a liquid hydrogen shortage one day. Again, the public information (below) does not do credit to the technology. We should build a real spaceplane someday. Hypersonics remains the part of my career where the more I learn, the less I know.

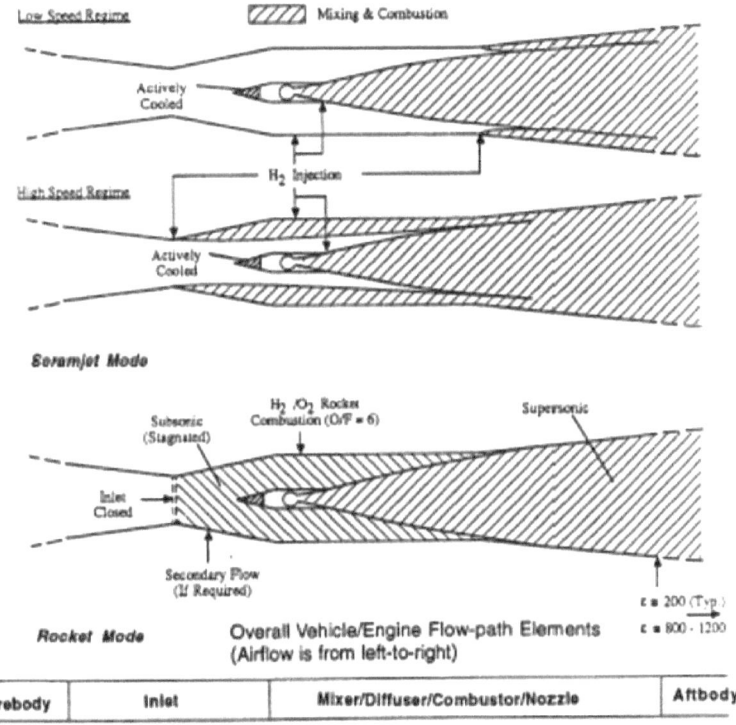

Cross sections of the flow path (only) of an air-breathing engine. From "The Synerjet Engine," selected papers by William J.D. Escher. Bill Escher, and Frederick S. Billig were foremost experts in the field of high-speed air-breathing propulsion.

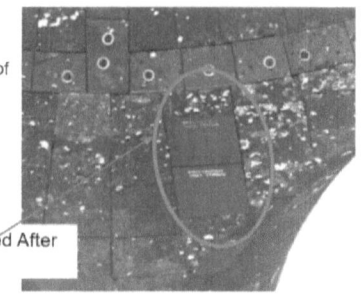

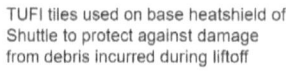
TUFI tiles used on base heatshield of Shuttle to protect against damage from debris incurred during liftoff

TUFI/AETB-8 Tiles Undamaged After Three Flights

Thermal Protection Materials and Systems: Past and Future by Sylvia M. Johnson, NASA Ames Research Center.

Yet technology around the corner was not the only place this happened, discovering a system where we thought we merely had a part. There are times we could have upgraded the Shuttles, only to find that changing a part affected too much else. Next-generation thermal protection tiles worked wonderfully when placed here and there on orbiters, with a world of difference from the original damage-prone tiles. But putting this new tile all over proved a leap too far, too late. The Shuttle's parts were notoriously intertwined to the point where the tiniest change seemed to reach out and touch someone, then everyone.

Systems are everywhere, as so much is connected or should be. In these pandemic days, I see more connections we might not otherwise appreciate. Perhaps because, at the end of the day, we are all connected as well.

THE VALLEY OF DEATH

We knew the valley of death was up ahead, as we had been there many times before. Most wouldn't make it. Well, to be truthful, we knew nearly none would make it. Wild ideas, new technology, and all those exciting, innovative projects that got anywhere from a bit of seed funding to serious dollars would all have to pass through.

Death Valley National Park as seen from the International Space Station. NASA.

In NASA-speak, technology has readiness levels. The levels go from one to nine. One is the kernel of an idea, a basic principle, while nine is flight-proven. New projects needing technology want to improve yet steer away from anything new, anything not at least at level seven, which has already been demonstrated and is close to the real thing. However, a teammate recently observed that some projects crave technology at level twelve, obsolete.

The valley of death came around level four when the gadget had to leave the lab. A tad over-dramatic, maybe. Yet the analogy truthfully captured the difficulty projects faced leaving the nest and fending for themselves in the real world. It was a desert out there, the heat, the vultures, and no water. For a dose of perspective, or at least sympathy, if you had spent 10 years of your career on a technology and then the plan to put it on the Shuttle went up in smoke, it was easy to feel you had wandered into the valley of death. *And* you knew you were unlikely to reach the next watering hole. Not to say the funding always went away, or your job. For most, it meant merely going back to the lab. Still, more years could go by, trying repeatedly to get that next big step approved. You might even get to the billion-dollar stage, around level six, only then to join the skeletons in the sand. Now, fifteen or twenty years had flown by.

This last month, NASA wrapped up testing on the James Webb Space Telescope (JWST). To provide some perspective about time flying by, I retired from NASA after 32 years, having begun my career there in 1988. If you began your career with the earliest studies about what comes after the Hubble space telescope, you arrived only a few years after me. You could spend your entire professional career on the James Webb project.

The James Webb Space Telescope. Image: NASA.

In dollars, the $10 billion (so far) on the James Webb is about half of what NASA has spent (so far) on the Space Launch System. It is also about half of what NASA has spent (so far) on the Orion crew spacecraft. This does not include the European and Canadian contributions to the telescope. More so, the $10 billion might be thought relatively reasonable. As the saying goes in NASA project circles, a billion here, a billion there, and before you know it, you have real money. Yet this is not about a dollar amount so much as a trend.

Fewer survivors are going through the valley of death each time. Each gets much more expensive, but budgets don't get much more dollars. Inevitably, fewer survivors have the funds to pass level seven, regardless of merit.

These all seemed really clever observations at the time. Increasingly expensive technology, fixed budgets, and here we all go - into the valley of death. Denying anything was amiss, the conversation also veered into how the valley was part of life, natural in a research organization. A healthy research organization must have many more misses versus hits. But that, too, denies the trend, the fewer hits every year. Removing the NASA-speak, this is another way of arriving at an economist's notions of diminishing returns and low-hanging fruit. In the private sector, it's about arriving at the proverbial question – does it scale?

If a satellite could see the impassable features from travelers' tales to the valley, it might see technological stagnation. Recently, you can see the notion in physics, the lack of progress, as the 40 billion dollar collider is probably not forthcoming. Having picked the low-hanging fruit, we can't just say "Next time, go bigger." In the past, you might go bigger and explore more places, but now it's about limits and fewer places. That's not to say there is a consensus on technology stagnation as a phenomenon. Critics of the notion of technology stagnation have a conversation like Calvin and Hobbes, where we are "not sure people have the brains to manage the technology they've got." By this view, we already have god-like technology (if paleolithic brains). Yet the case is often made that technology advances more slowly of late regarding real impacts on real lives. Or, as many professionals note, it costs much more every day for the newly minted medical doctor to put up a shingle even as lifespans stagnate.

As I often saw in complex projects at NASA, the answer may be in-between these views. And sideways. It's not stagnation, but definitely not God-like either. This brings us back to the James Webb. A telescope gathering light is not like flowing electrons in transistors. Moore's law does not apply. Twice a telescope diameter gets you four times as much light and more complexity. The James Webb will deploy sun shields, a central tower, a secondary mirror, and a primary mirror, among other housekeeping (deploying antennas, radiators, and more). All this will occur at a location beyond the Moon.

The telescope after James Webb can follow a straight line, just going bigger. Simply put, more area means more light. Imagine how soon we get to that next telescope, and for what cost? Expectations are tricky, and technology expectations are the trickiest.

I have the privilege of supporting the NASA Innovative Advanced Concepts (NIAC) program as external council. The recent portfolio includes exciting ideas for how our next-generation telescopes might come about. A low-frequency radio telescope on the Moon? Build it from what's there. ("FarView - An In Situ Manufactured Lunar Far Side Radio Observatory"). Another idea is to deploy a reflector one kilometer in diameter inside a lunar crater, packaged as a mesh. ("Lunar Crater Radio Telescope (LCRT) on the Far-Side of the Moon"). Looking for small asteroids, the most plentiful and valuable for resources, and the easiest to get to? Combine new information processing capabilities with multiple small telescopes. ("Sutter Ultra: Breakthrough Space Telescope for Prospecting Asteroids").

It was decades ago, a conference on hypersonics. An entire day was dedicated to wild ideas, and there was no lack of these. There were plasma deflectors to reduce shock waves on spaceplanes, laser craft, so huge ground lasers, and super-conducting innards in engines. Somehow, the program manager even finagled some funding for fusion energy (by asking if it fit into an engine). These were days when even trying to figure out bending time and space was on the table.

Today, there is similarly no lack of wild ideas. The kids really are, alright. Also, there is no lack of experts and speakers and motivational optimists ready to repeat a mantra about innovation. (If anything, these are more abundant than ever.) Connect the decades though, and perhaps a lack of innovation is not the root problem. Recognizing the trend set by programs that simply go bigger, as Augustine noted, the single airplane eventually shared by the Air Force and the Navy seems too possible.

More innovations need to make it past the valley of death, not fewer every year. Inversely, that means fewer eggs in the same basket, so there are actually funds available. Figuring out how to do this may be the best innovation of them all.

NATURAL AND ARTIFICIAL FLAVORS ADDED

The label read, "Natural and Artificial Flavors Added." So, I put it back. Artificial, we've been told, is just not good for you. We are almost at the same place with artificial intelligence. Alarm bells go off there as well, except in the form of Nobel laureates prognosticating about the dangers of A.I. There are also the occasional cautionary tweets on this from the billionaire who builds Starships and electric cars.

This is especially odd, as A.I.s are already everywhere. We have A.I.s in the justice system advising judges, in medicine translating protein sequences into images, and soon enough designing chips. I have a primitive A.I. called Alexa on my desk as I write, with another in the kitchen. I'm waiting for the day when I ask Alexa, "Are you sentient yet" and I get the answer, "did you call me primitive?"

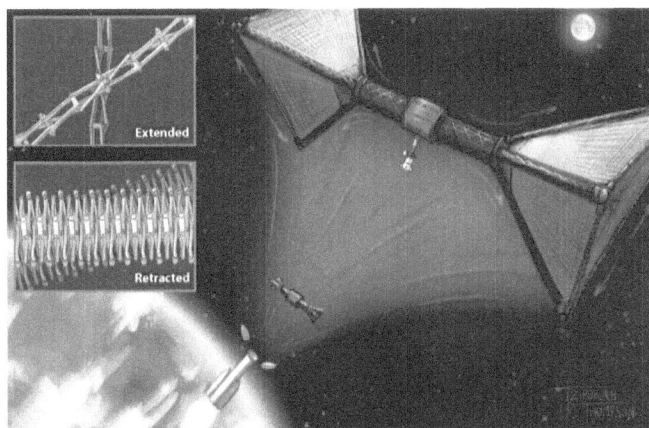
Kilometer-Scale Space Structures from a Single Launch, Zachary Manchester, Carnegie Mellon University.

The first A.I. most people befriended was probably Google, not even knowing this was an A.I. Like Yellow #5 it's just there, and no one notices it anymore. Yet mixed with all the warnings on A.I., there is so much promise.

It's easy to imagine an A.I. taking the protein crystals grown in space and revealing their structure, enabling a vaccine sooner rather than later. I've written before about my employing a primitive A.I. to explore the design space for reusable launch vehicles.

And then there is artificial gravity. Somehow gravity being artificial is never a problem. In sci-fi, it's a common trope, a necessity to avoid hanging your cast from cables all day. The dialog around artificial gravity is more convincing every day. Mention how any sufficiently advanced society will have technology that looks like magic. Add in some pop-sci with some real-sci and we can imagine the floor of a spaceship (or "down") coursing with some field we learned to control, emitting gravitons at will. But, first, we have to find gravitons, if they exist, and figure out how to create them artificially. That is, without the mass of a planet beneath our feet.

As artificial as that sounds, given the effects of space travel on the human body, "*artificial*" gravity sounds excellent.

Adverse health effects from an extended lack of gravity are too often under-appreciated in our quest to explore. There's muscle atrophy, mitochondrial dysfunction, and even liver stress of a level that might cause genetic changes. Long-term, your vision may never be quite the same ever again.

Here is where "artificial" is good for you. "Artificial Gravity Added." Grab that box! Sadly, NASA is not doing much in the field of artificial gravity. In NASA's defense, our mastery over gravitons is at least a few generations away (if they exist). Closer around the corner, a spacecraft would use a large enough structure to spin, the force outward acting as artificial gravity. Think of how a bucket full of water, spun on a rope fast enough, holds water in, even up-side-down. Or reminisce about the spinning room in 2001: A Space Odyssey, Dave Bowman doing laps round and round. (In this single continuous scene, director Stanley Kubrick includes an artificial intelligence, HAL, artificial gravity by means of a rotating section of the spaceship, and torpor, astronauts asleep for the long journey in space.)

I have the privilege of providing external council on some NASA research - the "NIAC." NIAC is part of NASA's really long-term thinking. The portfolio has a project on large structures for artificial gravity. So here you first work on the technology and challenges of a large space structure, then something large enough to spin would follow.

There are alternatives, of course, and no one knows how our exploration beyond Earth will pan out, or even if. For example, we could get to Mars using nuclear propulsion, avoiding some of the adverse effects of a lack of gravity by shortening trip times. Either way, any solutions have to figure out how to shield the crew from radiation, having left the relative safety of Earth orbit and its protective magnetic field. Space is a cruel, cold place, in no way obligated to us.

There is another concept NIAC has on our plate right now, making soil. All-natural, 100% old-fashioned soil. If you are onboard the Starship Discovery, a replicator will make that salad for you (and the plate, fork, and napkin). The replicated greens are nutritious, maybe even more than the real thing, Starfleet scientists being pretty good at this. (Except for alcohol, but that appears to be a policy choice.)

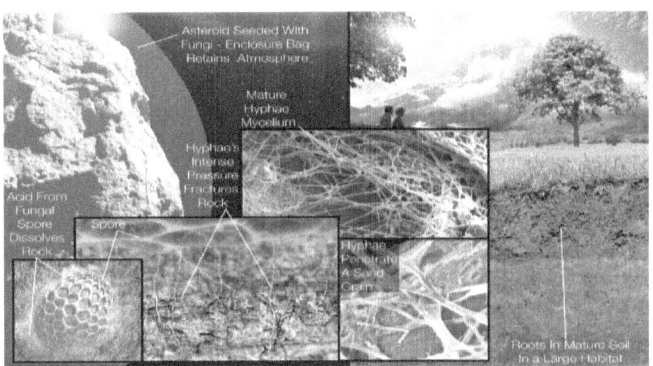

Making Soil for Space Habitats by Seeding Asteroids with Fungi, Jane Shevtsov, Trans Astronautica Corporation.

Closer to our time, real-live soil will be among the most precious commodities for people in space, alongside water, propellant, and that spare CO2 filter. Here, instead of taking all the soil you need with you, you take fungi instead. Mixed with lunar regolith and nutrients, the fungi will create living soil.

Meals Ready to Eat: Expedition 44 Crew Members Sample Leafy Greens Grown on Space Station, Image: NASA.

Imagine making our own soil in space. How wonderful

would it be if we could avoid a vast amount of mass to launch and deliver to the Moon or Mars as well as well? Again, no one knows how our exploration beyond Earth will pan out. But we won't get there by avoiding the hard questions, the real challenges of space travel, for one-day living years or a lifetime in space.

Some of what happens will seem artificial, whether it's an A.I. managing space traffic, advising on a spacecraft's health, or artificial gravity. We may also find a natural solution is the most elegant, looking beyond a brute force approach and engineering to biology and life sciences. Wild ideas are wild until they are the baseline. Natural and Artificial flavors could be just what's needed.

"IN THE END..."

Once again, it was that meeting, the one with the drone of presenter, questions, or occasional speech posing as a question, set to repeat mode. On cue, add the light-hearted tangent every forty-two minutes. We would cover issues and risks and endless lists, the former already happening, the latter possible. Also, as expected, the wrap included a pep talk by the manager in charge. The more problems, the more pep. Perhaps also sensing a repetition over decades, this one manager went with a classic - a fusion of "not our first rodeo" with "all is well." It went something like this – "The space station also had bad days, but in the end, time passed. We persevered and here we are – we have a space station. And all will be fine here too, in the end."

In the end. The words hung out there. In the end. I suspect these words took up rent-free residence in a few minds, mine at least. That is to judge from the reinvigorated chatter after the meeting, now reviewing the pep talk too. In the end.

The mirror of the James Webb Space Telescope vs. the Hubble's. NASA.

The call to persevere took the air out of the room. It was as if the airlock decompressed without time for me to grab a spacesuit. It came across that what problems we had spent months or years on were neither important nor urgent. The encouragement told us to accept progress was on our side, with a large dollop of inevitability. This should be a comfort as we toiled on our tasks. Indeed, if the attempt was only encouragement, it did not appear to rest on a belief about what we would soon fix or change, but instead that we just persevere. This was ironic, as "in the end" in economics is a call for urgency, since, in the end, we are all dead – so important matters should be fixed here and now (and paid for later, not in reverse).

It's not hard to run into the idea of inevitable progress – its realness surrounds us everywhere. Today, we walk around with tricorders that inform us if there are life signs ahead, or at least what's trending, and who just said what craziness. These are also our cameras now, digital where once we had to carry separate cameras with film. Film, the original expendable one-shot deal is long gone. Now we carry a reusable digital film of sorts. For storing those memories the substrate is no longer paper; it's electrons in the cloud. At least our moments can be uploaded if not us (yet).

In my lifetime, I saw large roll film and flashbulbs (in the hands of adults), then a work-horse 35mm camera, a novelty Polaroid instant, and too many digital cameras to count. Worth coming back to for lessons in technology, a camera's lifespan appears to decrease as we ditch expendable film. Today the cell phone works for most occasions (except extra special events, which call for a real camera.) And here we are, instantly exchanging pictures, where before we had to wait to see who had their eyes closed.

Long ago, I picked up a large format Yashica LM for a dime (LM for its "light meter," which still works), quite fascinated by the mechanical beast. I promptly relegated it to a shelf for show. Years later, my curiosity sparked, and with too much time on my hands, it turned out the camera worked wonderfully. It produced some beautiful, if prohibitively expensive, pictures. In this view, progress would seem just so inevitable. Nostalgia can't beat out new technology, even as we keep some relics around and dust them off now and then. Or is there more happening here?

A new height of technology in picture taking is coming around in astronomy, the launch of the NASA James Webb Space Telescope just days away. Just look at the JWST compared to its predecessor, the Hubble Space Telescope. In the JWST, we have a 6.5-meter diameter primary mirror, whereas Hubble's is 2.4 meters. Whereas Hubble can see nearly as far back as the first galaxies forming, the JWST will see as far back as the first stars. The JWST will collect 6.25 times more precious photons than the Hubble by surface area. Now we might follow up with a view on that habitable zone around Alpha-Centauri (and I can't wait to get the Robinson's on their way.)

Even with all these delays and unexpected costs, the JWST promises to make up for it all. History is poised to repeat the same way Hubble overcame its problems, even its blurry vision due to its mirror being ground just slightly off. On costs, the James Webb takes these leaps in scientific performance at about the initial cost of Hubble (inflation adjusted, minus the Shuttle launches). This is all measurable, undeniable progress.

This packed ESA/Hubble picture showcases the galaxy cluster ACO S 295, as well as a crowd of background galaxies and foreground stars. NASA.

And yet, "*In essence, NASA had to mortgage future high priority missions and research to address JWST's additional resource needs.*" This is from a recent GAO report looking at what we've learned from the JWST, necessary for what comes next. Using a common mortgage analogy, the wording here is rare for those not regular readers of GAO (or CBO, IG, and other independent reviews). These are reports with words like "challenge" used liberally (16 times in this one), losing the sense of import or urgency the higher the usage. Use the word "challenge" often enough, and before you know it, of course, it's space exploration - what challenges, what an adventure! Instead, laying bare that context matters is new and significant. A single house can show progress, but the neighborhood also matters.

Still, assuming all goes well, never a given, it will be easy to forget the "challenges." The sun-shields will deploy, forward and back, the tower for the primary mirror will extend, and the sun-shields layers will unfurl. Antennae and booms and gadgets will activate, along a journey to "L2" a million and a half km away, to a stable orbit beyond the Moon. There's a long list of what must fold out, extend, activate and awaken, and as complicated as it may sound, suffice it to say it's more than the lens cover on Hubble opening or solar panels pointing at the sun. Some progress comes with complexity, from scaling up but still having to fit on a rocket - origami style – and wanting to do more and better.

So, as I often saw in NASA projects, you can only scale up so far. Bigger is better until it's not. The project that survives, resolving its issues, and err...say again "challenges" is fortunate enough to have lived to tell the tale. The projects not so lucky won't get to pass along their lessons much at all, lacking resources and a pulse. Yet somehow, what worked and what didn't must inform the following steps – as difficult as it may be to put success in context or a failure on the radar.

The Very Large Array, where each antenna is 25 meters (82 feet) in diameter, but their data is combined electronically to give the resolution of an antenna 36 km (22 miles) across, and the sensitivity of a dish 130 meters (422 feet) in diameter. Image: NASA

These lessons are not new, about adding up, keeping projects in their context as part of a bigger neighborhood, or the sustainable limits of scaling up. The experiences on the side of the rocket and the ride appear not that different on the side of the payloads that do the science. For astronomy, one path ahead taking a cue from Hubble and the JWST could be fleets of smaller telescopes. Imagine something akin to the Very Large Array, only made up of many small telescopes in space in formation. Or as well, to get more photons, and see better, farther, for less cost, means innovations we can't yet imagine – but we urgently must. We will always ask more and more every day of our technology, and that includes that it does more and costs ever less over time. We take the pictures down to zero cost each, and the camera disappears as well – and still go for the upgrade. Perhaps the only progress that's inevitable comes with a good idea, and an ability to learn, improve and shift when needed, to see and go further, in the end.

IS THIS NOW?

Saying the universe is vast is an understatement, though it sounds better than saying we have no idea about the nature or extent of everything we have no idea about. These are not your project's unknown unknowns. This is where words fail. Recently, NASA revealed the first images from its James Webb Space Telescope. If ever there were a moment to bring out words like vast, immeasurable, or incomprehensible, this is it. Sagan put it best, referring to our exploration of all that's out there with his usual poetic perfection, where "we have waded a little out to sea, enough to dampen our toes or, at most, wet our ankles." Alternately, the blunt analogy would be that we are the baby who's been shown a quick peek of the night sky full of stars, on a small screen, in a Disney movie.

Stephan's Quintet, taken by the NASA Webb Telescope. NASA.

Like a new camera, the world quickly compared pictures from Webb with those from Hubble. As if we now have the latest full-frame camera hanging off our neck, the Hubble shots seem to be from the previous model. Though those old pictures do the trick and amaze us still, for now. Just as quickly, Webb's unexpected costs, delays, and the effect on other NASA projects, or non-projects as these never came about due to lack of funding, are staged to be forgotten. All is forgiven. This is the "Hubble Psychology," though it may be that soon we give joint credit to the syndrome and call it the "Hubble/Webb Psychology." In this syndrome, we finally see the benefit from an unexpectedly lengthy and costly project, and not just any benefit, but a striking success that overshadows all previous narratives, including the naysayers.

It's easy for us humans to embrace stories of sacrifice, the drawn-out battle for a just cause, worthwhile once victory is in hand (queue inspirational music in the background.) We are perhaps hard-wired this way, groups with this leaning carrying on, and their genes, if not specific individuals. This is not a simple trick to pull off. Plenty of large-scale projects offer sacrifices at the altar of progress only to find that when completed and ready for action, the gods are less than impressed.

Aside from these intangibles, some things can be counted. Surprisingly, the cost of building Hubble and Webb, adjusted for inflation with some arguable puts and takes, do not differ all that much. But like the new camera for a little more, the one you get today is much better. Perhaps there is FOMO in astronomy too?

Progress isn't always easy to measure, in dollars, sensitivity, or size of a primary mirror. Growth, as in improved performance, looked at in isolation will always seem like progress, but it's also important to ask what's sustainable and what's the environment. The fungus duplicating itself daily in a bottle is prospering for a long while, but when the bottle is half full, it's only a day away from discovering the limits of growth. Even when we know this, humans being more capable than a fungus, measuring progress is problematic.

Suppose one day we tackle the goal of going out there ourselves, mapping stars and studying nebulae first-hand. We could end up at the debate in Cixin Liu's "The Three-Body Problem," where we dream of people generations hence leaving our solar system and we research "curvature propulsion."

> *Cheng Xin pondered the strand of hair that had been moved two centimetres by curving space. "You are saying that you've invented gunpowder and managed to make a firecracker, but the ultimate goal is to make a space rocket. A thousand years may separate those two achievements.*
>
> *"Your analogy is flawed," Bi Yunfeng said, "We have invented the equation relating energy to mass, and we've discovered the principle of radioactivity. The ultimate goal is to make the atom bomb. Only a few decades divide those two achievements."*
>
> -CIXIN LIU, DEATHS END, BOOK 3 OF THE THREE-BODY PROBLEM TRILOGY

Lost in the Webb pictures, who wouldn't dream of one day going out there, a thousand years from now, or not as far? NASA once funded such research, up through about 2002, looking into the principles that might take space explorers well beyond anything imagined today. In a public forum, a NASA spokesperson, asked what NASA was doing about developing "warp speed," had a ready answer – "we are working on it." Today that's not the case.

Yet it's not only unimaginable distances to the places that have taken shape and detail since the first telescopes. It's also about time. As we look at the Webb images, we must ask, "is this now?" like an awakened host from Westworld. The answer is "no." This is not now. As with Hubble's, the pictures from Webb are everything as it once was, a very long time ago. We can look at them like vintage Super-8 movies, wondering where these people ended up. If we headed toward the galaxy cluster SMACS 0723, which we see in Webb's pictures at a distance of 4.6 billion light years, we see how *and where* it was. We picture the ball in flight. If we head that way one day, we would have to course correct to have enough energy to catch up to where it has moved since, away from us, even further.

Telescopes from Galileo's first models to Hubble to Webb remind us of all that awaits. Unfortunately, we won't be going there anytime soon, but we can dream. And thank you, NASA. Those pictures will do quite well for now.

Galaxy Cluster SMACS 0723, taken by the NASA Webb Telescope. NASA.

R&D INVESTMENT AND "HOW" – THE FINAL FRONTIER

"That's not gonna work." The person seated next to me mumbled this my way. It was many years ago, but not the only time, and the many versions of this story share the same look and feel. The presenter is up-front, enthusiastic. Some audience members lean forward for the Q&A. Fortunately, more often than not, the expert seeing a show-stopper rephrases that first impression. Tell us about this obscure point in the physics, or that phenomenon with proteins, optics, or the barrier that has kept the technology stuck in neutral. Then, a creative suggestion. Lastly, what are you doing differently?

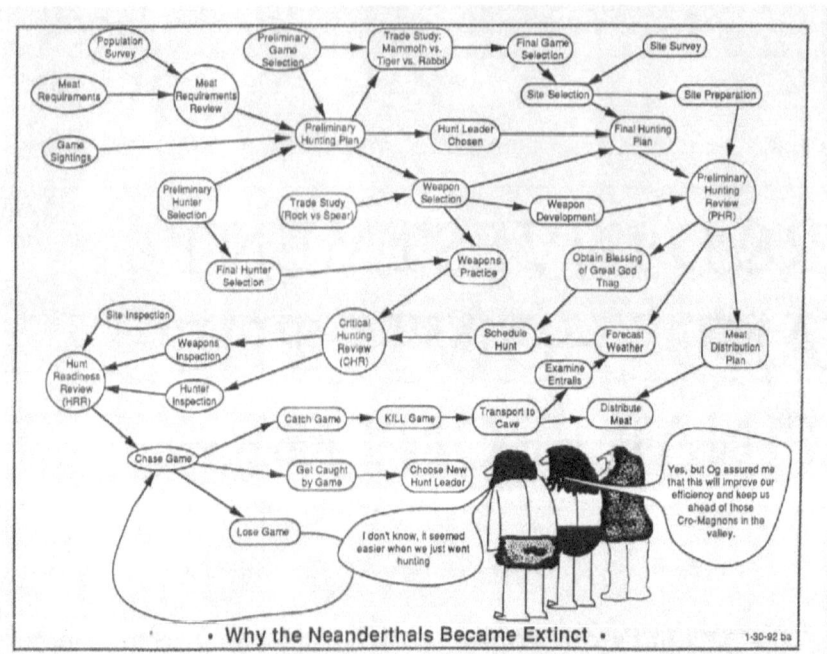

Too much process? Passed around in the 1990s in NASA. Unknown source.

Past this gauntlet, having been on the side giving and receiving, eventually, people make decisions. Bets are placed. Public or private, R&D, wild ideas and dreams are funded, with little mystery about why we play the game. We have to. The possibilities can't be ignored.

The mystery lies elsewhere.

The connection between ground-floor R&D and successful outcomes is poorly understood. The brainstorms carry us away. I've focused this way, too, and seen others dive deep, avoiding the distraction of far-off hoops to jump through or realities nearby. On the other side of the table, if you've been on enough committees, boards, and teams gathering ideas, you've seen plenty of process. Too much. It's not a process of looking at connections between a portfolio from long ago and big projects today. Rather, it's about due dates, credentials, and report templates.

There are some meager attempts at gathering up lessons

as tales of shortcomings. The titles could be "Adventures of an R&D Hit-man" or "Leadership Secrets of an R&D Hun." We learn that research dollars yield about as much return as sticking money in the bank – a poorly run bank. No discussion here goes for long without mention of big pharma, where 90% of new drug development fails. There are the new cliches, "fail fast and cheap." Old enough – whip out an older cliché, "stay hungry." Older still, say "out of the box." Like with quantum mechanics, it's easy to learn just enough to make conversation at a party. When Christenson's work comes along, and someone says "disruptive technology," this is the sign the party is over. We can all go home now.

Along too come the simulations, showing R&D to be like throwing darts or buying lottery tickets, except different. The sense we fight battles to win the war lasts until we compare time frames, returns, and everything else. Eroom's Law, "Moore's" Law spelled backwards, has the data to back it up, showing for the longest while we have been paying more and more for less and less. Per billion dollars of R&D we get fewer and fewer new drugs every year. Lipitor appears to have made up for all the losers. But then we must ask if marketing and patents and prices for one drug based on the failures are a plan for innovating.

Over too many projects, technologies, upgrades, and people with fantastic know-how, the structure of our R&D investment resists exploration. This is the rooftop of a building with a curious lack of connection to its surroundings.

Here, humility occasionally makes an appearance. We quit believing that we "know how to pick 'em" like some Wall Street guru who beat the market five years in a row. Nassim Taleb reminds us that randomness will fool us every chance it gets. We always believe we will be the exception, courtesy of our biases and mental shortcuts. Mull things over endlessly, and the tiger has us for lunch. Throw in some quick but erroneous logic, and at least we ran the one time that mattered if also (wastefully) all the others. Add a dash of optimism so we go out and about the next day, none the worse for wear. A short memory comes in handy, too.

Naturally, technology is physical, but methods are abstract. You can touch the gadget, and even specific enough ideas seem clear. Even the most rarefied challenge, say getting to Alpha Centauri sooner rather than later, is tangible compared to pondering *how* best to invest to get such leaps and bounds.

So why ask why? Why wonder how to innovate how we innovate? Are we those Neanderthals looking at the new hunt plan?

Pay-to-play will appear to make sense if you've been in any federal agency long enough doing R&D. Except it may be blindly assuming the same ample resources from better times. Keep calm and carry on. Too often to surprise anymore, the scenario plays out as if forgetting it worked well in the past only because resources were more abundant. It's no wonder you met some success as you wandered before. You wandered so much. Recently, in a reminder the seven fat cows are followed by thin ones, the budget for NASA's Space Technology Mission Directorate got a haircut. Add in inflation and how R&D can easily exceed the inflation rates seen on tables about consumer goods, and it's far worse.

As if fewer resources aren't bad enough, R&D investment gives in to the temptation to ignore the disclaimer in the prospectus. We expect past returns to be indicative of future performance. Why mess with a good thing? Being odd-man out, it's not a stretch to ask when the law of diminishing returns rears its ugly head. Any return we got years ago for our seed corn will taper off. It's not if, it's when. Repeating how we succeeded before is a guarantee we go no further.

Ramblings about R&D must come to the valley of death – as if it's not enough to dwell on diminished resources and declining returns. So much research goes well, but suddenly, it's too expensive for everyone to graduate. This is the infamous "valley of death," where failing cheaply is no longer allowed. Still young, the technology middle years are carefully curated and managed – fulfilling the prophecy that "really proving it works is expensive." Predictably, big projects given the go-ahead can't fill their shopping carts with what were merely dreams twenty-five years ago. They reach for older than that. The shelves for what's recent yet ready are bare.

I'm told that spreading knowledge is an essential R&D function in the larger scheme of things, a surrogate to address scarcer resources and slower returns. Overcoming denial, partnerships spread knowledge, admitting you must get others laboring when your resources are limited. The commercial shift also avoids the diminishing returns of repetition. As with many partnerships, the NASA Commercial Lunar Payload Services, failed and tipsy landers and all, is pushing into this territory. Yet this is more about the valley than the R&D in its earliest stages. Figuring out new ways to do what was done before is not the same as figuring out new things.

The temptation at the end may be to shrug. There will always be voices peddling new metrics, interdisciplinary this or AI that, to address how NASA or anyone performs R&D. Keep at it. Or the contrarian view where "we already have *too much* process." No need to worry, as shifts happen organically, arising from the melee. More NASA projects are favoring connecting innovation to projects through partnerships. And challenging projects like the James Webb Space Telescope were once R&D-ish, an immature idea about a very large mirror that folds to fit in your rocket. But these remain a variation on a theme where we hope enough wandering makes some R&D real, one day, in some project.

How do we *increasingly* connect our NASA R&D to successful projects and programs? Some fads, some sayings, informal notions things are fine, or better years lie ahead, with quips about someone's secret sauce? Well, that's not gonna work.

THE NUTS AND BOLTS VS. NASA BUDGETS

There is the micro and the macro, the one down at the nuts and bolts, hardware I would see up close and lay my hands on, the other a view from 100,000 feet. Zoomed in, nose at the nitty gritty, there's a drawing, a specification, forces, and dimensions. It was common in Space Shuttle operations to find myself trying to make heads or tails of a design that pre-dated my arrival by a decade at least. What were they thinking? Eventually, it made sense, most of the time, not always. With these shims and the spring-loaded washers, the Shuttle Orbiter and its external tank are held together with a specific force, or otherwise, a bad day. But our measurements show something is off, by a few thousandths here, or maybe there? So not enough force holding it all together.

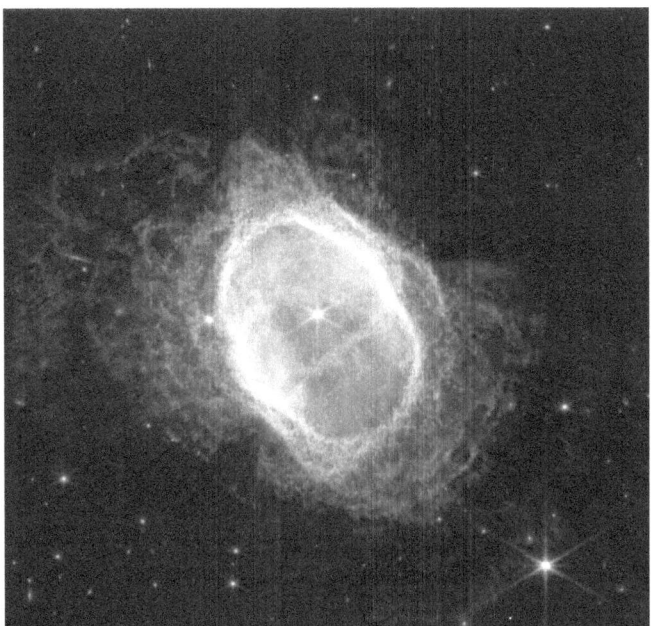

The Southern Ring planetary nebula. One of the first images from the James Webb Space Telescope. NASA.

Later that afternoon, I switched gears and zoomed out to a wide-angle view, analysis for some future project or technology. Here bracketing an answer was give or take a billion. This was often about direction, which way to head. This meant the macro-view was about choices, not being sure of the specifics but making a case for why one approach was preferable to another.

Jumping back and forth between these worlds was not common. Nevertheless, I believed the mix was necessary, the one lending reality and experience to the other. Eventually, I had to choose which world to work in full time (spoiler alert, I went into a world of advanced projects.) Which zone you are in is not easily confused, most of the time. But like much in life, I found there was an exception.

❋ ❋ ❋

With the fantastic first pictures from the James Webb Space Telescope, the narrative of a project long delayed and costing many times more than advertised initially seems poised to be forgotten. Instead, the new story is about success, science, and the splendor of the starry night. Norm Augustine, a faithful steward of the aerospace enterprise, recently said, "overruns in cost and schedule can often be forgiven through great accomplishments, and the Webb telescope is exhibit one." Forgiveness is sure to follow amazing feats of engineering that open the door to unimaginable scientific knowledge. Yet right along, this confuses the micro with the macro. A data point is not a trend, and criticism of a trend is not criticism of a project.

If you haven't heard of Augustine's Laws in the aerospace business, you have not yet attended your first meeting where a project gets "re-baselined." Here we find out the initial cost and schedule were not official. As well, somehow, no one knows who put out that cost estimate, and in either case, the manager says the numbers should never have been put out there in the first place. This new estimate will be the real one. This is not the re-baseline you are looking for, though. There will be another, then another. Occasionally a voice in the desert will holler out how none of the cost or schedule estimates ever much passed a sniff test. This is because the project has been doing R&D for years. They just didn't want to admit it. The project's first years gathered knowledge, from which there eventually came better cost and schedule estimates. Then we spot a trend among these projects, and someone says, "looks like Augustine's law."

After plotting many projects, Augustine notoriously stated what he saw as the trend for US fighter jets in his typical folksy way. As each fighter jet gets more expensive than the last, and fewer of these are produced, you arrive at Augustine's 16th Law:

"In the year 2054, the entire defense budget will purchase just one tactical aircraft. This aircraft will have to be shared by the Air Force and Navy 3½ days each per week except for leap year, when it will be made available to the Marines for the extra day."

Yet such a trend includes trusty and venerable aircraft like the F-16, a success story and a remarkable jet that will be flying with upgrades for the foreseeable future. This reminds us analysts, a trend is something else entirely from a specific project. While it's tempting to say trends are many projects together on some graph, that also misses the point.

When I wore my program analysis hat, I found making sense of the tension between a project and a trend meant a mental shift. It wasn't only the cliché of stepping back from the trees to see the forest. Instead, I had to also understand the environment. For NASA, the environment is the yearly budget. The connection between projects and budgets is a matter of actual hardware, people, and challenges caught up in an environment where the sand will shift under their feet.

Norm Augustine also notes this in his recent interview – noting a presumed (and oddly precise) $3 billion shortfall between NASA's content and its yearly budget. Putting aside the off-the-cuff number (first seen in the wild over a decade ago), NASA's budget trend is simply not good. NASA's purchase power, its dollars after adjusting for inflation, is 20% less than in 1995. If NASA had $5 back then, now it has $4.

This means three paths for any NASA project, do better, do less, or after projects get racked and stacked, do fewer. First, your project might do for $4 what once took $5. Second, you might tell the boss, sorry, this next probe does less than the last one (good luck getting funded.) Or lastly, where once you had ten probes on the burners, now you have eight. And this is the good news.

Coming out of the pandemic, supply chains are not quite what they once were, not that excess capacity was ever admired. Russia invading Ukraine has sent oil and gas prices up into a tailspin and likely up again soon. Yet, for odd reasons, the US Federal Reserve thinks increasing interest rates will help cool off inflation, usually a result of easy money and so excessive demand, not broken supply. At any rate, higher interest on the US federal debt will pressure the entire US budget, as interest on the debt is particularly sensitive to the cost of borrowing. More interest on the US federal debt squeezes other spending. Other spending – as in NASA.

Would a scaled-up Webb, the LUVOIR space telescope at 8 to 15 meters in diameter vs. Webb's 6.5, thrive in such an environment? NASA's recent decadal survey admits the answer is probably not, merely on cost. Now imagine where *probably not* was the optimistic answer.

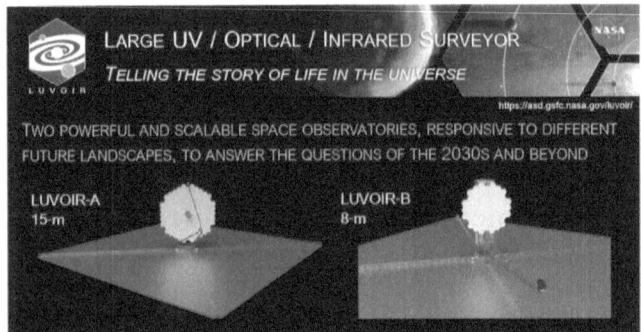

A proposed space telescope to come after the James Webb Space Telescope. The larger LUVOIR-A at 15 meters or the LUVOIR-B at 8-meter diameter would be larger than the 6.5-meter JWST or the 2.4-meter mirror on the Hubble space telescope. NASA, Mission Concept Study Final Report, LUVOIR.

❉ ❉ ❉

Apart from a particular project, its merits, and the scientific wonders it promises to deliver, an analyst must step back. Way back. Farther. Another Law: "Time screws up everything." (Also 2nd Law.) Everything might, with optimistic assumptions, right now, maybe add up (in theory).

Program analysts know it's when we plot against time that we find the real problems no one wants to hear about. On decadal time scales, a program analyst at 100,000 feet quickly sees a chapter for the promised report. It's conceivable we will find ourselves in the year 2030-something, glad to have upgraded Hubble way back when and again. The Webb has recently ceased functioning, never designed to be refueled or upgraded like Hubble. So then, there is no Webb follow-up in the foreseeable budget environment. Worried about an ISS gap? What about a space telescope gap.

This is all very linear, and thankfully time and space are not always linear. So rather than doing less, or doing fewer, picking "do better" is a possibility. Because NASA enjoys doing the impossible, imagine where we "potentially produce larger lenses at a fraction of their current price and in a fraction of the current manufacturing time." As far out as a "liquid lens" may seem, this type of R&D responds to the environment ahead.

Similarly, innovative exoplanet science, characterizing planets in other solar systems, may come through starshades. Here, specially shaped shades in space suppress a star's light allowing us to see the nearby planet. NASA's Innovative Advanced Concepts (NIAC) program is also funding a look at a starshade.

NASA budgets are limited, and prospects look worse. What's not limited is the ability to imagine doing better and more, for less. We have seen NASA embrace commercial partnerships that do exactly this. The practice of placing wonderful projects on the chopping block, so the fewer surviving projects can do wonders, will hardly pass. Neither can NASA lose sight of the trends, budgets and the dull stuff setting the environment, and how to prosper anyway. Multitask, and you can do this as you take in the pictures that leave you in awe.

A space telescope lens to dwarf all telescope lenses, via innovative technology. "FLUTE," Fluidic Telescope Experiment, Edward Balaban, NASA Ames Research Center.

RISING WAGES, MEET TECHNOLOGY ADOPTION

Our space sector does not lack news about new tech, business deals, or novel things to come. But, with so much happening, imagine for a moment that the nature of the churn also changed. Would anyone notice? With too much noise, do we miss changes in the signal? The usual tropes marry change and technology talk around NASA, a scientist, a discovery, SpaceX, a start-up, a medical breakthrough. The day's technology lies along a spectrum, from the hubris of Dr. Frankenstein bringing to life his creation to the humble sense of tech as merely a tool. Out of nowhere come the instruments of change, affecting everyone, carried along, or overrun, as "it lives!"

It's easy to see technology first, then the result. For engineers and scientists, it's how we were raised. The prequel doesn't get the same press. Yet we know our environment encourages innovation or not, and once in a while, it changes the rate of change.

A while back, in May of 2021, The Conference Board published a look ahead at the US labor market with a curious reveal. The working-age population in the US isn't growing and won't be growing much anytime soon. The last part is the interesting bit – the US labor force will not grow much *in the coming decade.* This report likely got lost in the fray at the time. We had other things to worry about, like when we would get our next shot or return to the office and if our routines would ever be the same again.

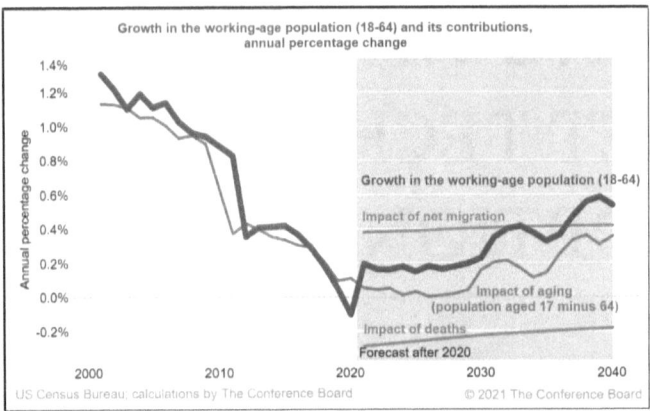

Growth in the working-age population, by The Conference Board, May 2021.

Now nearly two years later, we see rising wages near the top of the screen, in the headlines over and over. Alongside, there's more about low unemployment, today's 3.4% unemployment rate was last seen the year Armstrong walked on the Moon. Right nearby is the US Federal Reserve steadily increasing interest rates, so stories about inflation and housing. The reserve intends to dampen high inflation (mainly not due to rising wages.) When the toolbox only has an interest rate hammer (and a few odd tools with acronyms like QE), guess what tool gets whipped out. Eyes watering over yet?

A short-term view says this too will pass. We just had a global pandemic, with millions of deaths globally, 1.1 million in the US and counting. Then there's long-Covid. Never mind, higher wages will swell the workforce back to size. Also, people were forced to leave jobs in one place for another, but all this will settle down. Or it's immigration that's down, but higher wages will (again) make workers appear (from somewhere unspecified.) These takes comfort – be patient, we'll get *back to normal,* whatever that was, soon.

The longer view is quite different. If demographics are destiny, the number of people working is what it is and won't go up soon because the past has already set the future in stone. As more people leave their work years behind (myself included), the number of people around to fill the empty spots is, at best, forming a very short line.

Imagine the Conference Board's longer-term scenario proves correct - a US workforce that is what it is for the foreseeable future. Here, demand for labor way above supply for a long time means suddenly, easily justifying technology rather than technology struggling to come to life. The sense of "waiting it out" and not investing in new tech while the labor squeeze goes away just won't seem credible.

That presentation to the boss about the new machine, where the numbers were hard-pressed to "close"? Your small fan club for that new technology will soon see a boom in membership.

It's for someone else, perhaps an economist, to write about technology adoption *speeding up* when the demand for labor *far* exceeds the supply. (It may be a story about farmers who faced with fewer workers more readily adopted tractors and combines. Or it may be about the adoption of automation in the US in the 1950s.) What I can share is my experience with advanced technology in the US space sector.

In the Shuttle program, working in the world of "upgrades" and new technology meant learning a hard lesson over time – failure was the default option. Billions were spent analyzing, testing, or prototyping new technology, but little made its way onto the Space Shuttle. We could take solace, saying at least we learned a lot and look at everything we put on the shelf. If not Shuttle, "someone will run with it, one day."

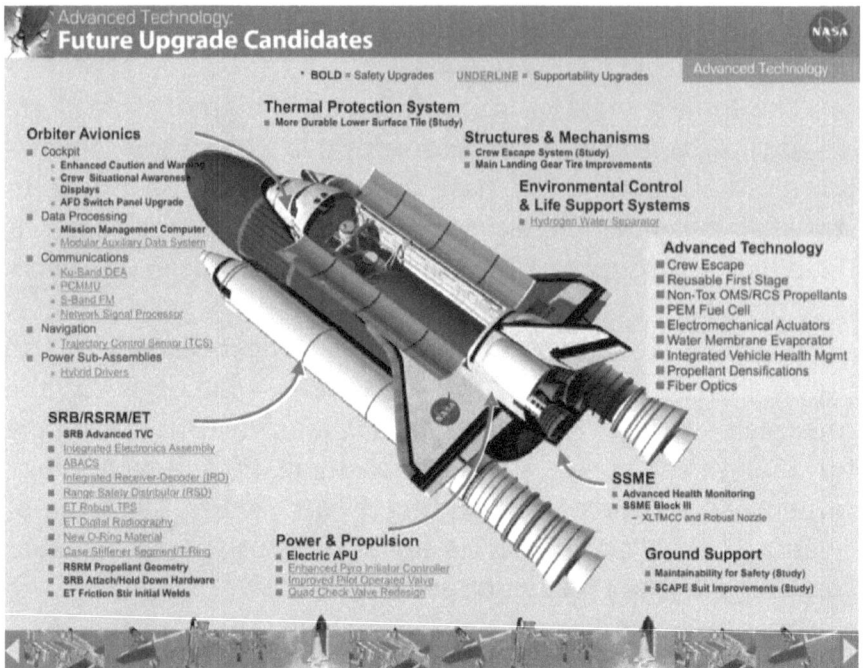

Space Shuttle Future Upgrade Candidates, circa 2000. NASA

If you wanted to replace leaky, problem-prone hydraulic actuators on the Shuttle with electric ones, you might feel heartened by rumors the SpaceX Starship has gone electric. (We eventually concluded the best of all worlds was hybrid – some electro-hydrostatic actuators and some pure electric here and there. Don't get obsessed with consistency.) This was one of many technologies that fought to be put on the Shuttle, or on a next generation system. While the Shuttle had toxic fluids, we fought for non-toxic replacements. For the Shuttle's 1st generation tile, people tried to justify 2nd and 3rd generation tile and materials far more resistant to damage. (These made their way onto the rear at the engines.)

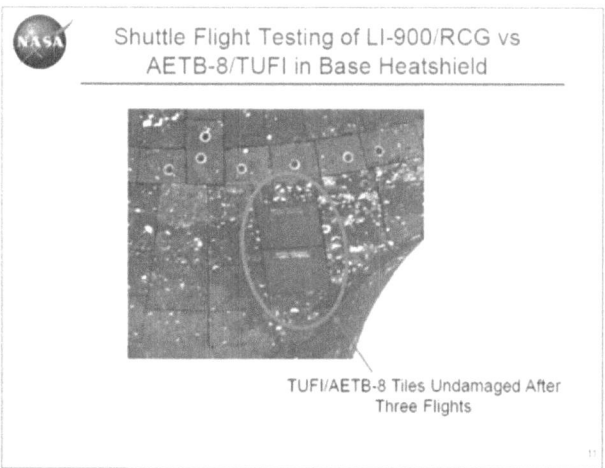

What a difference a little technology can make. Notice all the damage in the older, original Shuttle tiles with the white chips all over vs. the two next generation tiles. "Thermal Protection Materials: Development, Characterization and Evaluation," Sylvia M. Johnson, 2012, Ames Research Center, NASA.

We once joked that new technology would get onto the Shuttle when the last person who understood the old system died. NASA spent billions of dollars in *attempts* to upgrade the launch control centers. When a last push to get new computers finally made some headway, I recall being told the last person at Kennedy who knew how to fix the old-Apollo style keyboards had retired. Now we seriously had to get this done!

* * *

As unavoidable as gravity, selling advanced technology got married to reusability and flying more often. What better way to justify the price tag for a new gadget. Each cheaper flight had a savings from the new technology, so the more you flew, the faster you could recover an initial investment. But NASA has only so many payloads, so we had wandered into new territory – "who needs to fly so often?" Where was the demand (or any pressure) for more launches that would justify the automation, technology, the "new and improved"? Sounds familiar?

A leads to B leads to C – and so new technology led to a high flight rate at low cost which led to reusability and growing commercial operations. Yet as my days in the Shuttle program became a full-time job with advanced projects, technical talk gave way to non-technical debate. Beyond volume, for us flight rate, or payloads meaning markets, we always circled back to incentives.

I've often written about incentives in aerospace projects. Incentives define the environment in which ideas, like seeds, try and grow. An environment encouraging a high speed is one thing. Encouraging acceleration is probably quite another.

Fast forward, and now we see a robotic apple picker that would be at home in any sci-fi movie, and we would think it's just special effects. In aerospace, we have robots printing your rocket or winding fiber if it's composite. This may seem a long-winded version of the obvious, more of what anyone already learned in 3rd grade - "necessity is the mother of invention." What's different now is necessity appears poised to increase.

Time will tell if persistently higher wages mean even faster technology adoption – or for some, technology adoption *finally*. Though there will always be the temptation to find cheaper labor elsewhere, even in aerospace. NASA-speak flipped out the word "mitigate" as if it was always Option A ... and B and C.

For spaceflight, whether launchers or what gets launched, low volume presents a peculiar barrier to new, more productive technology – *not needed here*. Yet what goes on outside aerospace, where ever higher volumes of production are a goal, will not happen in a vacuum. Once you breathe life into a new technology, don't be surprised it wanders to unexpected places. Even when we saw a technology couldn't find any takers, at least we too wandered to unexpected places, learning and growing. So, suppose high wages do mean embracing new technology, faster, with less hemming and hawing. In my time at NASA, we saw enough technology hurtles, and failings, and systemic issues we slowly learned dwarfed all that. We also knew when to say, about some challenges – *these are the problems we want to have.*

BREAKING THE SPEED OF ANALOGIES

Analogies. Everyone loves a good analogy, all the better when they cut right into the heart of a matter. Our space biz is not immune to the allure of analogies, chock full of complex backstories, technology, and eccentricities just begging to be simplified. Though when it's oh so clear, it's probably oversimplified. Elsewhere, the analogies are sounder, that is, until someone says "Moore's law," and the laughter says we just stretched that analogy until it snapped. One analogy still making regularly scheduled appearances compares rockets and space exploration to airplanes and air travel. Or rather, we see our rapid progress in the air as compared to the slow progress, or spurts and backsteps, when reaching for orbit and beyond.

A still of some Super-8 footage taken by my father out the window of the jet in 1968, and my first time on an airplane. I was three years old.

The air to space analogy is ironic given air travel in recent decades. The sense of progress in the air no longer seems so sure or obvious. Boom Supersonic announced a $60M agreement with the US Air Force last week to research a supersonic airliner. For comparison, the first "A" in NASA, for aeronautics, had an $827M budget last year, or 3.6% of NASA's budget. The "S" for space in NASA gets nearly the entirety of the budget, as probes and robots in space or for human spaceflight. This split seems natural if we assume a mature US air sector already gets all the investment it needs, public or private, because of so much progress. It would seemingly appear most of the job in aeronautics is done. We easily explain NASA's relatively paltry aeronautics budget if we believe all there is to research is mainly in the past. That is until we see a company saying there is plenty more to come, and we can yet again go faster like the Concorde did, only better and profitably as well. Going further, if we are so enamored of the analogy of air travel to space travel, what happens if we run with it – at the risk of stretching it till it breaks?

The backstory is a classic, as a couple of high-school

drop-outs turned bicycle makers fly the first powered, heavier than air airplane in 1903. The Wright brothers put the government effort to shame (that spent 70 times as much.) The next part of the story usually skips to the DC-3, a mere generation and a half later, making it possible for airlines to fly passengers profitably. Skip again, so the analogy continues, to the hugely successful Boeing 707 jetliner in 1954, providing an experience that anyone today would find familiar – minus an infotainment system. Finally, we have the massive 747 in 1969, the same year as we first land on the Moon, and in 1976 people are zipping around the globe supersonically in the British and French Concordes.

My first flight on a jet was in 1968, and I was three. Ironically, though I did not know it at the time, I would never fly as fast ever again. All the jets started slowing down soon afterward for the sake of fuel efficiency. But that is just the start of the air-to-space analogy going sideways.

In 2003 the Concorde was retired, never recovering from the 2000 Air France crash leaving Paris. I recall landing in New York in 1999, and the sight of a Concorde on the runway nearby was enough reason for our pilot to make an announcement. He could have said – *"if you look out the window on your left, there's a plane I would like to be flying, instead of this bus."* Coincidentally, I had just recently read "The Concorde Story" some months earlier. Intrigued, dial-up line and all, I had even checked for a ticket – New York to London, $10,380.78. And yes, that's one way.

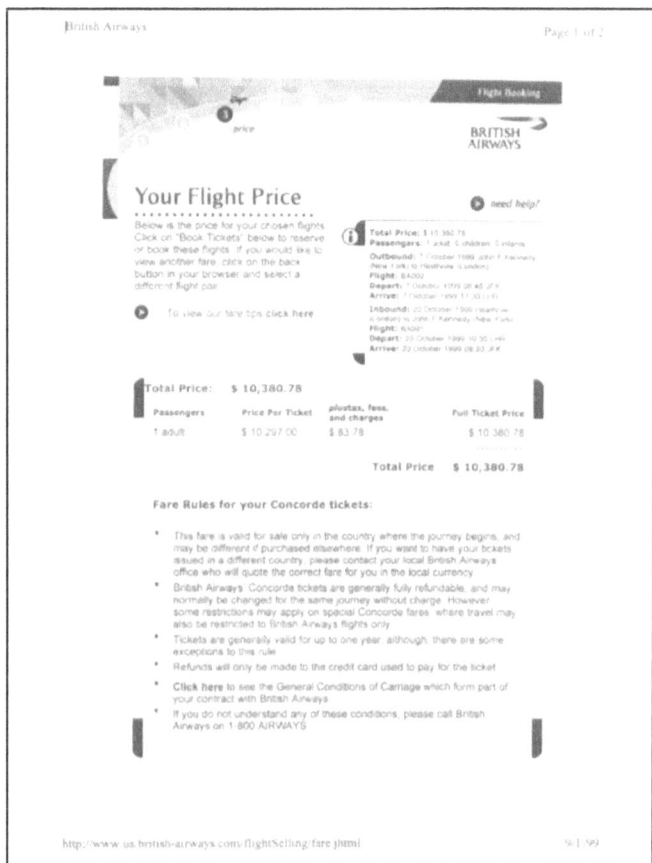

A ticket quote for the Concorde in 1999 - $10,380.78, one way, NY JFK to London Heathrow.

By then, I was fortunate to have worked on assorted spaceplane projects in NASA – in between my Shuttle duties - as far as projects that were very hardware poor can be called projects. It was then I had the pleasure of meeting people who had worked on engines in the 1960s that would be an aeronautics major's dream today. Bill Escher wrote the book on technologies combining air and rocket modes – "The Synerjet Engine." Fred Billig helpfully pointed out key concepts. *"This is how that works - for real."*

Meanwhile, engineers from the National Aerospace Plane (NASP) days in the early 80s gave me first-hand accounts of just how the "Orient Express" program fell apart. NASA would continue with the X-43 a while, with the DOD running with the X-51 after that. As some NASA engineers kept at it, all that remained, often supporting the US DOD, was studies. But these seemed more about preserving and passing along knowledge, not generating more. The running joke was, "I'd tell you all about it, but then I'd have to shoot you." This was not a formula for expanding knowledge, and as useful as these studies were, rehashing is no replacement for creating.

The NASA flyer for the X-30 National Aero-Space Plane Program. NASA.

The NASA flyer for the X-30 National Aero-Space Plane Program. NASA.

So, we first had Kitty Hawk in 1903. Just 52 years later, we have people flying in jets to see their relatives, seal a deal, or just to vacation. A couple of generations, then a little more, and we go even faster, supersonically, as if a natural next step. Here we get the sense so much has happened in aviation in so little time, from invention to democratization. We have flown the friendly skies and pushed boundaries further, faster, all well within a human lifetime.

Human spaceflight might mark its Kitty Hawk moment in 1961, with Gagarin and Shepard. Now 61 years on, we are hardly near when the family is packing to visit relatives on the Moon. (Great - now we have Auntie on the Moon – that's better than when she was right near Disney!) Of course, it helped that air travel already had destinations we knew were ready to receive us with all the amenities - from Paris to Fiji. More so, the analogy breaks down when we look at the state of air travel today, where the case against advances like supersonic travel is about time. Going faster can hardly matter much when we are on the ground for most of our travel time. To reflect on spaceflight's challenges and progress (or not), we're going to need a bigger analogy.

NASA is building the X-59 supersonic demonstration vehicle to prove out quiet supersonic flight over land. The technology would "reduce loud sonic booms to a quiet thump." NASA.

Now is when someone who is not an engineer aptly says it seems we have more technology now than we can manage. The crash of the Concorde in 2000, then September 11, 2001, and the loss of Columbia in 2003 reset much about the world. Yet I have cause for optimism. We still have reusable launch vehicles, now Falcon 9 boosters, elegantly landing vertically. Its legs open moments before landing like some sci-fi flick we might have found overly imaginative once upon a time. And now we may have a go at it again in supersonic flight, with Boom Supersonic, among others – including NASA. All of which takes us to the better analogy between progress in air travel and spaceflight, the desire to explore. As Carl Sagan said, *"For all its material advantages, the sedentary life has left us edgy, unfulfilled. Even after 400 generations in villages and cities, we haven't forgotten. The open road still softly calls, like a nearly forgotten song of childhood. We invest far-off places with a certain romance."*

The analogies never end, to guide and to question. The US government used Air Mail to encourage air transportation, and not so long-ago NASA used cargo to the ISS to promote commercial launch services. NASA even used the Air Mail analogy to justify the commercial cargo approach. On the flipside, the questioning where spaceflight is said not to have advanced in its first few generations, as did air travel, neglects parts of each story. Spaceflight went from a man in orbit to men on the Moon in less than a decade. Yet we have not returned since. Similarly, air travel made great leaps up to the Concorde, only to backtrack as well.

As helpful as comparisons are, analogies and all, perhaps the real kinship between air travel and spaceflight is in the challenges. It's a few steps forward and two steps back for journeys we have to hope always leave room to go further.

OF STARSHIPS AND SPACEPLANES, AND ROADS LESS TRAVELED

The first Martian to visit Earth is a young boy, a twist on visiting the old country of your parents. To him, Earth is a distant place, with an odd blue sky, and sparking some anxiety. This was the premise of a short film at the Kennedy Space Center visitor complex in the early 1990s, "The Boy from Mars." The obligatory scenes on a ship from Mars include cheesy dialog matched only by an even more cheesy space cruiser. After, everyone must transfer to a spaceplane to get down to Earth. Creatively, rather than landing like an airplane, or Space Shuttles, the spaceplane's gentle glide back to Earth ends with a pitch-up maneuver and a vertical landing.

Hollywood loves landing winged spaceplanes vertically, with no runway required. After all, why would any director want to limit just where the action might go? So, of course, we get wings and things, a landing pad instead of a runway, and plenty of window seats.

Spaceplanes getting loaded with cargo, as envisioned in the Kennedy Space Center Vision Spaceport Partnership, 1998-2001.

A memorable scene follows the family's arrival at the spaceport. Reporters wait to receive the first Martian in what seems uncannily like a gate at Orlando International. That is, before 911 when waiting for your friends and family right outside the gate was the norm. After the long ride, everyone's a bit cranky, and the boy has acted up a bit. Millions of miles on the road will get you in a mood. Finally, the family exits the gate, both parents scolding the boy non-stop while he tries to get out an apology. The reporters never get a word in edgewise, left only to say pretty much *how some things never change*. The most exciting interplanetary voyage we might dream of today is just another long flight to Florida in the 2060s.

Spaceplanes got a lot of attention from NASA, not just Hollywood, in the following years. These concepts were air-breathers, true hypersonic spaceplanes that fly much like airliners. Taking in air along the way, they would not have to carry all that oxygen like rockets. At face value, this seems beyond obvious. Who needs to have heavy liquid oxygen tanks to get to orbit when there is air all around us? A spaceplane would meander a bit, going sideways in the atmosphere for longer rather than shooting straighter up like a rocket. All the better to grab all that free air and accelerate.

These are not idle concepts, more PowerPoint than real, or just studies. NASA and the US Department of Defense first poured serious money into these technologies with the National Aero-Space Plane in the early 80s. (By the time I joined these tasks, we would work NASP-Derived Vehicles, or NDVs, in NASA's penchant for acronyms that contain acronyms.) When it should have been front and center, unsaid in all this was a vision of space travel as rather pedestrian. At least as pedestrian as flying in a tube in the air can ever be, with all the wonder that should still amaze us if we paused for a moment. If Grandma is ever coming back from visiting her daughter and the grandkids on the Moon, the "flow-down" is spaceplanes.

Start with the vision, and where does it take you, *in reverse?* If we envision a family leaving the gate at the spaceport making a show for everyone to see, you have something like an airliner. Except being a spaceport, this airliner is a spaceliner, and it came from orbit.

Unfortunately, funding the necessary technology for spaceplanes came to a standstill by the late 1990s. Some insiders even questioned the whole idea. The definitive (public) compendium on air-breathing spaceplanes, "The Synerjet Engine," by Bill Escher of NASA, nonetheless shows none of this as ill-conceived, remote or impossible. Spaceplanes are not something for another generation. Instead, the vision remains much closer than anyone might think. The matter is about steps along the way, a very long but doable to-do list.

Rather, today we have SpaceX Starships, at the same time we have hypersonic missiles across the sky in Ukraine, and NASA going back to the Moon. This curious mix of news would seem unconvincing if it were the short description on the back cover of a sci-fi novel – set in the future. Yet here we are.

There remains a long list of items for NASA astronauts to step foot again on the Moon. Besides completing a crewed shakedown flight of the new SLS rocket and the Orion spacecraft, NASA must finish –

- a Gateway, a new lunar space station where crew will stay a while after arriving on Orion, but before leaving on a lunar lander,

- a new SpaceX Dragon XL spacecraft to ship supplies to the Gateway,

- new Spacesuits, with Axiom Space and Collins Aerospace,

- a SpaceX Starship, in its lunar lander form with legs,

- a SpaceX Starship, in the refueling tanker form, returning to Earth to be captured by the arms of "mechazilla" as it slows just above the landing site (which is also the launch pad)

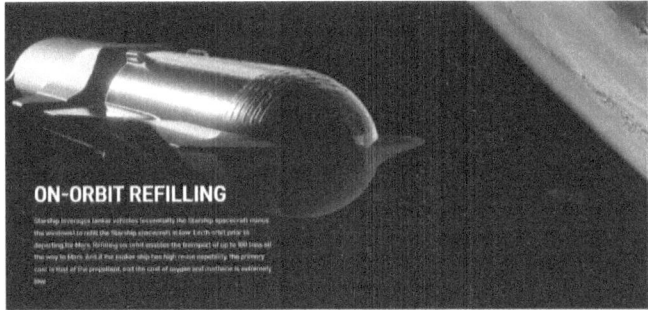

Tanker (far left) mating with a lunar lander (right) or another version of the SpaceX Starship. Credit: SpaceX.

Not on this to-do list - spaceplanes. Yet a vision of routine travel to and from space would seem to require returning to Earth comfortably. While a necessary step is reusability, as Starships will do, there remains a distance to go to coming back home as if stepping off an airplane and out the gate. After a lack of further interest by the late 90s, a resurrected interest and funding in hypersonic propulsion has come not from NASA but from the US DOD – and the secret world of weapons. Oddly, the last breath of interest in a reusable rocket with wings also came from the DOD, with the DARPA XSP. Yet these would never meet, the one-off to a world of closed doors and classified data, the other finding Boeing bowing out of the partnership.

Drawing up a list for true hypersonic propulsion (*which those merely ballistic Russian missiles are not*) would have different items on the to-do list –

- assorted new materials for propulsion and for thermal protection, (We often joked in our DOD projects how all we needed was "unobtanium")

- a new engine where combustion occurs at supersonic speeds (if only it were as easy as the cliché of lighting a match in a hurricane, as that would be low speed)

Seemingly an impossible list, NASA now has partners that can make the seemingly impossible happen. SpaceX, with the reusable Falcon 9, and the Dragon spacecraft, show NASA and the private sector can make lightning strike. No one would have predicted a reusable Starship as a NASA lunar lander as recently as a few years ago. Unobtanium jokes aside, thermal protection materials today may appear on the surface a lot like yesteryears, but these are not your parent's Space Shuttle tiles. Producing composites today is a very different matter too, no longer the craftwork of patient technicians and engineers mired in goo, but one of robots and better quality. (Boeing and the DARPA XSP showed the way here.)

Neither are the software and simulations today even remotely like their ancestors from the 90s. That would be an insult, like comparing an AI model today to my first spreadsheet from 1986. Instead, what we often found to be a barrier in our DOD projects was not physics so much as a lack of funding and incentives. And it was never a world of funding or asking for a blank check.

Connecting all these to-do lists could be seen as the missing piece, more of that cliché of productivity improvements lagging behind technological advances. It takes a while to figure out how all the new gizmos and gadgets work together. We already had wi-fi, the chips, touch-screens, and apps. All we needed was someone (actually a thousand someone's) putting it all together the right way, in that first iPhone.

It's easy to say *join the to-do lists the right way*. Lightning must strike twice in the same place, and close in time, for transport to and from space. So one day, sooner rather than later, the easy flight from orbit is memorable only for the scene that family made at the gate.

* * *

Part 4

PART 4 STORIES OF THE ONCE AND WONDROUS SPACE SHUTTLE

My first intimate experience with the Space Shuttle came before I was with NASA. The loss of the Space Shuttle Challenger and her crew in 1986 is the kind of event I mark time by. There is before, and there is after. In the lobby of my engineering department, the following year, a documentary about the tragic event played on loop all week. My mind played games because it seemed every time I passed by between classes, the video was at "Challenger, go with throttle up." Two years later, I was employed at the NASA Kennedy Space Center. Over twelve years and about seventy-five launches, I was privileged to be a part of a grand undertaking with the most talented team in the world. During and after, I worked with teams on the technology side of the house to answer the following questions: What have we learned, and what comes next?

We admitted to ourselves a sense of urgency. None of the people I worked with were fooled into thinking we would never lose a Shuttle again. Contrary to the public advertising by NASA, we knew the Shuttle far too well. We should have answers ready. Sadly, that day came sooner than we expected.

The loss of Columbia and her crew changed the course for NASA for this generation. I would hear often afterward, "The Shuttle proved reuse too expensive." Or "It made no sense." These were some of the kinder words. Of course, this is not true. I continue to work with NASA teams. We remind ourselves the only real failure is trying and not learning anything.

THE FLOW MANAGERS GLOSSARY

It's a struggle not to write or talk in NASA-speak, a skill, and a bad habit, from years of intensive training. I named my writing space "Zapata Talks NASA." On the one hand, I would talk about NASA, but I would also offer a talent, how I spoke this foreign language, and my writing would attempt a translation to plain English. Here is how we laughed at ourselves, aware of our obscure lingo in the early 1990s Space Shuttle program.

```
                        FLOW MANAGERS GLOSSARY


WE ARE BASICALLY CLOSE TO SCHEDULE  = WE ARE BEHIND SCHEDULE

WE ARE ASSESSING XXXXX = WE ARE FURTHER BEHIND

WE NEED TO COME TO GRIPS WITH XXXX = WE ARE EVEN FURTHER BEHIND

WE NEED TO UNDERSTAND THE IMPACT OF XXXX = WE ARE GOING TO SLIP BUT
                                           HAVE NO IDEA HOW MUCH

WE ARE EXTENDING THE END OF THE FLOW 2 WKS. = WE SLIPPED 2 WEEKS

EVERYTHING APPEARS TO BE READY FOR TOMORROW = I DON'T KNOW WHERE WE
                                              ARE BUT I'LL KNOW
                                              TOMORROW

IN ESSENCE WHAT HAPPENED IS... = I'll NOT SURE WHAT HAPPENED

EFFECTIVELY, YOUR ANALYSIS IS CORRECT = I DON'T KNOW BUT YOUR ANSWER
                                        SOUNDS GOOD TO ME

OUR NUMBERS DON'T AGREE WITH THE COMPUTER #'S = THE BEAN COUNT WENT UP

THE PAPER WAS LATE GETTING TO THE FLOOR = WE PUT IT ON THE SCHEDULE
                                          LAST NIGHT TO RUN TODAY

WE ADDED TIME TO THE SCHEDULE FOR... = WE DID NOT FINISH ON TIME
```

A flow manager was someone with the job of managing the endless scheduling and tasks for launching a Space Shuttle.

THE CASE OF THE $5,000 SOCKET

The socket cost $5,000. But we got a good deal for three at $15,000. This might sound like just another story about a $300 toilet seat, but there may be some rhyme or reason behind $300 toilet seats. Or even $10,000 toilet seat covers. If you make a plane and then one day no one makes more of those planes, and one day you need parts for that plane, like a toilet seat cover, don't be surprised finding parts is expensive, as you pretty much have to pay someone to make that part for you - from scratch. All this with lots of paperwork along the way to make everyone rest assured that it's the same thing as the original part. And all that paperwork is effort, and all that effort is money.

This socket had no such excuse.

But let's go back in time a few decades. The Space Shuttle is flying again after the loss of Challenger. It's the early 1990s, and among the items that keep getting much attention are the valves connecting the Space Shuttle orbiters (Discovery, Atlantis, Endeavour, and Columbia) and the External Tank with the propellant. The day a Space Shuttle crew was in town for a thank you and pep talk at KSC, I grabbed a picture of that valve for the crew to sign (yes, that's Story Musgrave and the crew of STS-44). It's not exactly the kind of valve you might think opens and closes a 17-inch line that is going to flow super-cold liquid hydrogen (pictured) or liquid oxygen at flow rates that "empty the typical backyard pool in a few seconds" as the KSC guides were fond of impressing on the tourists when spinning off numbers about the Shuttle's impressive feats.

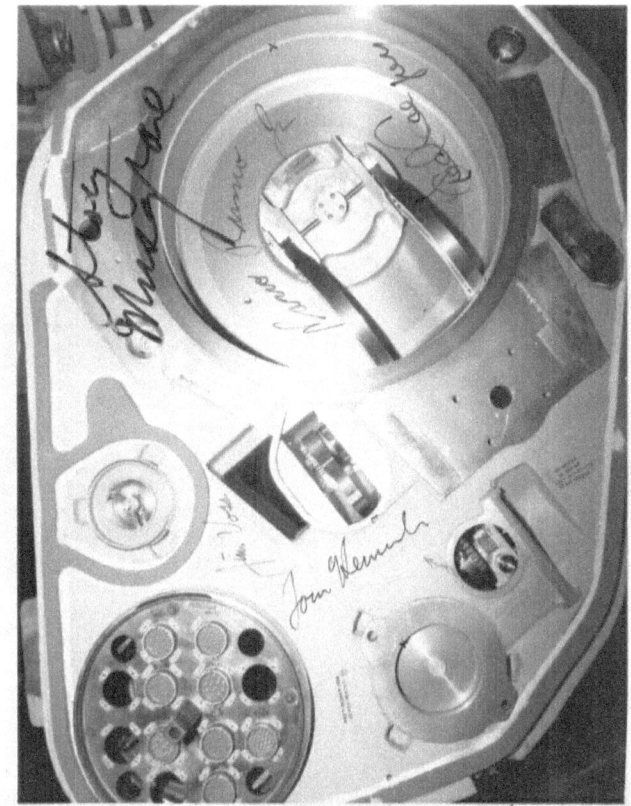

A Space Shuttle 17-inch umbilical valve, external tank hydrogen side, signed by the crew of STS-44, 1991.

Oddly, when this valve would open, it would remain in the middle of that hydrogen (or oxygen) flow. Again, if that has not sunk in, the big plate with the two arms you see in the picture would flip into *the middle* of the massive flow of cryogenic propellants. In the middle. In the way.

Of course, after Challenger, this was put on the list of the bad days in waiting. What if the propellant flying by the valve so fast applied so much force, this "flapper" pretty much a leaf in the wind, that it just shut? This would be catastrophic. Immediate disintegration. Now, the obvious thing was to change the design. But that begs the question of how the original design came about in the first place. The designers knew full well what they were doing – this was a lightweight design, *and if you are going to put something the size of a 737 into Earth orbit, every ounce* counts.

The Space Shuttle orbiter side 17-inch valve umbilical assembly circa 2005.

Alternatives to this valve would probably weigh more, and in either case, might also be forced closed by the tremendous flow of hydrogen and oxygen and take forever to retrofit. In the meantime, someone figured out you could put something between the existing valves – a latch – that ensured they remained open during the flight, mechanically restrained, if still in the middle of the flow.

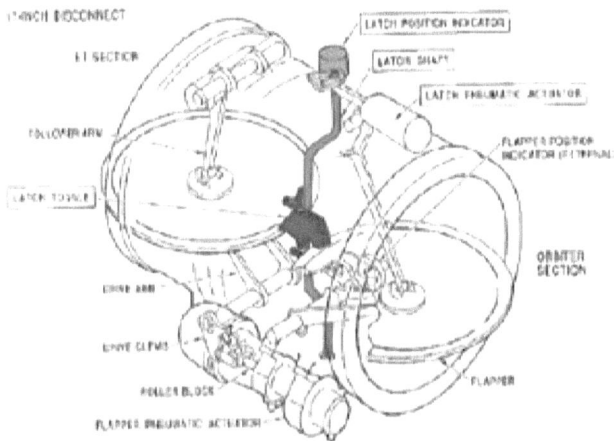

The latch (in red and blue) was added as a retrofit to ensure the 17-inch valves on the Shuttle would remain open in flight with propellant flowing by. Image: NASA.

And so, yes, there was still that pucker factor. And it was high. The latch was a bit of physical insurance, just not mentally. Think of aerodynamic forces on the wing of an airplane, flowing under and over the wing just right in flight. Now think about this valve in the middle of all this monster flow of super-cold *liquid* hydrogen or oxygen, creating hydrodynamic forces. All is fine as the liquid flows under and over the valve (the "flapper," as it was called). Except there's a problem. No one's flying this airplane – the plate must be at a certain angle to the flow so it's all in balance with the propellant flow. That is, so it does not get forced closed by the flow.

Additional insurance was simple. Once open, we would check the valves five ways to Sunday for all their settings, angles, and positions.

Sounded simple. Not quite.

It turns out this valve was never meant to be opened unless the External tank was mated to the orbiter. Except then, there was no way to see what was happening inside the valve assembly. Of course, engineers, like nature, find a way. The valve could be manually opened using a wrench and a socket reaching into a small space along the side of the contraption.

Except there was no socket the correct size.

If this story starts sounding like, "For lack of a nail, a horse was lost," - it sort of is. A train of events began with a design likely driven by a desire to weigh as little as possible, as far as later discussions could agree. It was an elegant solution in its own way. The designers knew exactly what they were doing. On top of that, you have to imagine a designer who could have gone, say, with a ½ inch shaft on the valve, so a ½ inch socket, but that would have been over-designed, a waste of mass. SO inefficient. Alternately, it could have been 3/8 inch, but the analysis of loads would have said that it was not sturdy enough. Naturally, the designer went *in between* those sizes.

Now anyone who has done a little handy work knows you can grab the socket that's a little larger, and it might do the job, except watch it – because you either lose a knuckle when it gives unexpectedly, or you will damage (by rounding) what you are trying to turn. Sure enough, that's what we did for some years.

Until someone had a bright idea – why not get the sockets that would be just right for opening and closing this particular valve as often as our hearts desired without risking damaging it (in my naivety at the time, I was part of this). That can't cost much, can it? Damage a multi-million-dollar valve, who knows, and besides, the right tool for the right job!

Then I saw the bill. The contractor (name withheld here to protect the innocent) had a socket made from scratch, just perfect for the job. It would be $5,000. But just in case, we ordered a few. So $15,000. When the sockets arrived, I held one the way you would a piece of expensive jewelry taken out from under the bright lights and glass for you to peruse.

For lack of a nail, a horse was lost. And so on and so on, but what was really happening? It's not really about mass savings. Not entirely. Today, we have come a long way as we fight the rocket equation, a seeming realization that hardware doesn't have to bow to the tyranny of the rocket equation. *Some mass might just get around that equation.*

Hardware is expected to be designed with affordable operations in mind. In a system like the reusable Falcon 9 booster, we see a massive (pun intended) shift to where the seeming philosophy is propellant is cheap, but effort is expensive, so why not just up-size? Then, that last ounce quickly loses importance in perspective. The rocket equation tyranny is still there. You just got a little creative about avoiding the bowing. And this is a path to preventing the sockets that damage the valve or the better ones for $5,000 apiece.

We faced similar questions in another reusable booster project - the DARPA XSP. There, we would try adding wings and things and returning a booster horizontally, like the Shuttle, under no power. Contrast this with the Falcon 9 reusable booster scaling up for extra propellant, re-igniting engines, and returning vertically. The balance is not apparent. We are still a way from where all boosters like airliners look alike, probably because the best solution or the solution that wins (which may not be the best) is still out there. How might we locate that design? We would have to balance all the variables, mass, design, cost, technology, practices, and more (with care to avoid the compromises that lead to $5,000 sockets). For that, we need some powerful thinking, human and otherwise.

The real challenge is locating where added mass adds value faster than losing payload, a benefit not from being absent but by what it does when it is present. Reusability, sustainability, reliability, and lower prices will follow when mass is seen for the friend it is, not someone to avoid.

IT AIN'T OVER TILL IT'S OVER

"It's over. They worked out an agreement."

The boss stuck his head into the conference room to make the announcement. Judging from the look on everyone's face, clearly, there was some confusion. Interrupting our meeting and just blurting out late-breaking news does this.

How could it be over when we were just getting started?

We met to discuss the improvements we wanted from a competition to hand over the Space Shuttle's operations to a single company. At the time, the Space Shuttle operation was made up of many contractors, pretty much those who built it. The Shuttle was not a business that took delivery and sent the manufacturers away to await calls for tech support. Instead, the manufacturers stuck around in a seamless segue to operate and launch it all.

Weapons category	Total U.S. contractors			Current U.S.-based prime contractors
	1990	1998	2020	
Tactical missiles	13	3	3	▸ Boeing ▸ Raytheon Technologies ▸ Lockheed Martin
Fixed-wing aircraft	8	3	3	▸ Boeing ▸ Northrup Grumman ▸ Lockheed Martin
Expendable launch vehicles	6	2	2	▸ Boeing ▸ Lockheed Martin

February 2022 "State of Competition within the Defense Industrial Base" – Note the use of the words "Expendable" and "vehicles" instead of launch services. Also, still thinking about "prime contractors" years after competitors like SpaceX have started launching national security space payloads.

Perhaps I finagled an invite to this meeting because of my interest in the possibilities. Like a kid in a candy store, I saw no end to the goodies we might get once "under new management." However, in retrospect, my invite was probably for lack of takers, with few others seeing much change ahead. Frankly, I landed in this meeting suitable for many pay grades above me as my naivety exceeded my curiosity.

It was 1994 or so, and the following year saw the formation of the United Space Alliance, a joint venture of Lockheed Martin and Rockwell (the latter really being Boeing at the time.) With time we would find that the Shuttle program was a Humpty-Dumpty who fell off a wall at birth and was quite tricky to put back together again. With time the expected road-shows advertised all the savings since USA was formed. Yet budgets (so costs) had already started dropping before the formation of USA. As well, adjusting for a reduced flight rate in the years following the reduced costs just ended up where they should have been – less cost for less launches. Right-sizing vs. improvements, and costs per year vs. cost per flight, were often conflated in the ensuing debates about savings. How could the output of a new, bigger, single contract not be significant "efficiencies" after all?

Just as the boss said, the scheme to compete who might best operate the Space Shuttle was over.

Competition, or better said, *lack thereof*, is the subject of a recent report from the US DOD, among all the aerospace news competing for our attention. There is not much new in this news, but there are plenty of oddities. For one, nine years after the DOD opened up national security launch to competition, and 5 years after SpaceX first launched a national security payload, the report lists only Boeing and Lockheed Martin as "primes" for "*expendable* launch vehicles." The words *reusable* and *SpaceX* appear zero times in the report. As if continuing the thought competition is good, perhaps, who knows – the report asks we forget pricing. The early line "*Although studies of this trend have not found a strong correlation between consolidation and increased program pricing...*" (my underline) is not a glowing endorsement of a need for competition in the US DOD.

Ronin, like myself, often worked on projects for DOD. (Full disclosure: I worked with DOD as a NASA employee, on and off since 2000, until I retired in 2021. If you want to understand the ins and outs of this, your next item to dissect is who runs what between Star Trek's Starfleet vs. the Federation.) And therein come to mind more terms missing in action in this DOD report on competition, and its seeming lack of enthusiasm for it, to judge from the lukewarm phrasing throughout. Federal agencies can all learn from each other, but NASA's successful commercial partnerships are never mentioned. That would mean mentioning Falcon 9s, Falcon Heavies, or Starships. It would mean talking about redundancy in partners for *next-level complex* systems, from Starliner, to Dragon to Cygnus, all for less costs (even combined) than having just a sole source. It would also mean mentioning fixed-price contracts and having two suppliers for basic operational needs, each a backup to the other. *Finally, it would mean touching on the ability to walk away from a capability by constantly nurturing more and better capabilities elsewhere in the wings (pun intended).* The alternative is the occasional anti-trust action preventing a merger. But mostly, we get a slow spiral toward ever fewer contractor options. If we do not put a trend in reverse, it will go forward.

Space is hard. But it's just physics. Competition is harder, as it's about people. A single, constant signal is so much easier than trying to listen to many voices. This is so even when those many voices point out the changes necessary to keep an organization relevant. In 2013 I had the rare opportunity to even enlist prime contractors as co-authors to espouse anti-trust measures that would make recent FTC actions seem bashful by comparison. (The meeting included a Risky Business moment of realization.) We knew then as now, in the annals of NASA and launch systems lessons learned, there's as much to add about new policy as about new technology.

This is not the first report ever on how consolidation in the US aerospace sector might not be good – at all. In 1996 GAO got the understatement of the year award for *"Documented savings from business combinations have not always been as great as initially expected."* And this is when mergers just began, with more to follow. Ten years after United Space Alliance, Boeing and Lockheed again joined forces for launching DOD payloads as United Launch Alliance. Ironically, it fell indirectly to NASA to add competition to the national security space launch picture by investing in SpaceX for cargo to the ISS. (After Starfleet vs. the Federation, let's also debate why all Starship construction is at Utopia Planitia Fleet Yards on Mars.)

All of this has happened before, and it will happen again. But does it have to? Or contrary to what the boss said, and more like Yogi Berra, "It ain't over till it's over."

MAKE GOOD CHOICES

Soon, NASA will load propellants onto its new Space Launch System – the "SLS." This test will span a few days, a whole shakedown and practice run, much like the launch countdown starting at T-72 hours for a Space Shuttle. This is an exciting moment, the end-to-end system seeing liquid hydrogen and oxygen for the first time. While a relatively uneventful few days are a possibility, experience says to expect the unexpected.

Hitting inconel, valves, aluminum, and a plumbers' nightmare of tubing, seals, and nuts and bolts with super-cold liquids brings to mind many sayings about cryogens. How do you load cryogens? Very. Slowly. What does it take to ignite hydrogen? You looked at it wrong. What does hydrogen do? Well, it just wants to get out.

For example, lines that might seem impervious, close cousins of an army tank, have been known to have too much stress. In 2002 a hydrogen vent line cracked, through and through. It simply cracked. Of course, these lines were old and had seen many a freezing moment, then back to hot Florida days. The vibration of many liftoffs had given them a beating over the years too. Hear that crack when you drop an ice cube in soda, and you get the picture.

A cracked hydrogen vent line on the Shuttle's mobile launcher platform. Image: NASA.

Thermal properties are likely high on the checklist going into loading super-cold propellants in a new and complex flight and ground system. The Space Shuttle dedicated books to expected temperatures across many situations for just about everywhere you could imagine. A "wet-dress rehearsal" will show just how close reality matches expectations – all those models and sims, and the component and sub-system tests.

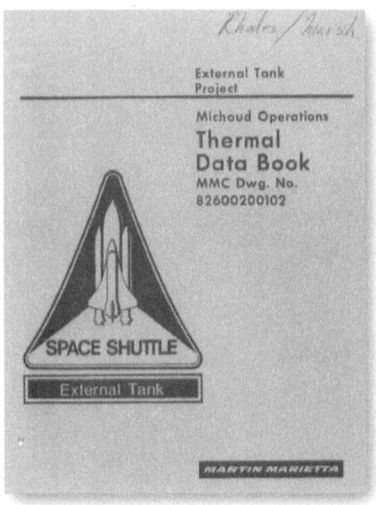

For the Shuttle, thermal data books analyzed expected temperatures across ground and flight systems as these see cryogens. Here, the cover of the Space Shuttle External Tank Thermal Data Book, by Martin Marietta.

For the Shuttle, propellant loading was naturally a cautious and slow affair. Given the pedigree of the SLS, in scale, propellants, and infrastructure, there will be similarities. Unfortunately, for this SLS test, the public will not have audio of the usual steps as they occur, chilling down facility lines, then flight propulsion, followed by a slow fill, before ramping it up to a fast fill. Sadly, we will miss that voice of NASA public affairs, calm and Zen-like, saying things like "T-minus 7-minutes and counting." Nothing is more peaceful than the predictable phrases during a NASA launch countdown. For SLS, perhaps next time.

6.2 Tank Loading Procedures
This section addresses the procedures and time lines for loading the LO2 and LH2 tanks.

6.2.1 LO2 Loading Procedures
The main time line events are as follows:
- Transfer line chill
- Orbiter MPS chilldown
- Slow fill to 2%
- Fast fill to 98%
- Topping to 100%
- Replenish
- Terminal countdown and liftoff

These seven events are elaborated upon further in this section. Figure 6.8 and 6.9 presents an ullage pressure profile during loading.

Transfer line chill
- Initiate SSME purging
- Turn on pre-press and-icing purge heater
- Open main-fill valve, replenish valve, TSM vent valve, TSM drain valve, ET vent valve and helium bubbling ΔP to between 1.0 and 1.9 psid

Orbiter MPS Chilldown
- Open TSM vent valve, TSM drain valve, main fill valve, and replenish valve and
- Change pump speed to 1000 +/- 100 RPM and operate pump at 1000 RPM for 2 minutes, then
- Change pump speed to 2850 +/- 100 RPM and operate pump at 2850 for 2 minutes
- When 2 minutes complete close main fill valve, set replenish valve at 70 percent, and close TSM drain valve
- When orbiter inlet pressure is greater than 34 psig, or skid outlet pressure is greater than 56 psig, close the replenish valve and allow feedline to drain through TSM vent and engine bleed valves
- When Orbiter inlet pressure drops below 10 psig and skid outlet pressure drops below 32 psig, or 12 minutes has elapsed since replenish valve was closed, open replenish valve to 100% open
- Perform facility line flush to opening TSM drain valve and transfer line main fill valve and flowing at 1000 GPM for 3 minutes
- After 3 minutes, close transfer line main fill valve and delay for 15 seconds before closing the TSM drain valve
- When Orbiter inlet pressure exceeds 10 psig, close the TSM vent valve and start LO2 fill to 2%

The Space Shuttle liquid oxygen loading sequence, 1 of 2.

LO2 Slow Fill to 2%
- Close TSM vent valve and start 11 minute timer
- 3 minutes later, verify fill rate between 250 and 290 gpm. If not, decrease 11 minute timer by 0.4 minutes for each 10 gpm above 290 gpm, or increase 11 minute timer by 0.6 minutes for each 10 gpm below 250 gpm
- At end of timer, end slow, fill and proceed to fast fill

LO2 Fast Fill to 98%
- Close LO2 vent valve and verify temperatures are within limits before proceeding
- Open main fill valve, close bypass valve, change pump speed to 3450 +/- 100 RPM
- Maintain ullage pressure by vent valve cycling between 2.2 and 8.0 psig

LO2 Topping to 100%
- Initiate topping when either 98% sensor is 50% wet. To do so:
  - Reduce the pump RPM to 2850 RPM
  - Open the vent valve
  - Open the bypass valve, open the replenish valve to 100% and verify the main fill valve is open

LO2 Replenish at 100%
- Initiate replenish when either one of the two 100 percent or (100(-) percent sensor is 10 percent wet. To do so:
- Close transfer line fill valve
- Reduce Replenish to 20 percent, modulate to maintain 100% sensor No. 1 15 to 75% wet
- Reduce pump speed to 2700 +/- 100 RPM

Terminal Count Sequence and Lift-off
When drainback is to start:
- Close replenish valve,
- Close inboard fill and drain valve (PV10),
- Open the TSM vent valve and TSM drain valves,
- Close the ET vent valve, and
- Turn off the helium bubbling

Pre-press tank:
- Turn off anti-icing purge at T-2M50S
- Pre-Pressurize with Helium to 20-22 PSIG at T-2M35S
- Retract GO2 vent hood between T-2M50S and T-0M20S
- At T-19S close outboard fill and drain (PV9)

6-18

The Space Shuttle liquid oxygen loading sequence, 2 of 2.

6.2.2 LH2 Loading Procedures

The main time line events are as follows:
- Facility/Orbiter chilldown
- Slow fill to ECO sensors
- Fast fill to 98%
- Topping to 100%
- Replenish
- Terminate replenish and terminate sequence

This procedure to control tank pressure during slow fill to ECO sensors, and fast fill to 98%, is to enhance the TPS bypressurizing to pressures above those expected in flight. When the TPS transitions through the glassification temperatures during fill, the flight pressures the TPS experiences will not exceed proof factor requirements. The six events are elaborated upon further in this section. Figure 6.10 and 6.11 presents an ullage pressure profile during loading.

Facility/Orbiter Chill-Down
- Vent down from 30 psig in the transfer line for 1 minute
- Re-pressurize line from outboard fill to skid valves to 6 psig
- Vent ET for 105 seconds
- Open transfer line, chilldown, and main fill valves, allowing unpressurized flow into the transfer line
- After 4 minutes close the transfer line valve

LH2 Slow Fill to 2%
When last chill-down step is complete:
- Start pressurizing storage tank
- Close ET vent valve and pressurize to 46.7 psia
- Vent valve cycle using 43.7 psia and 46.7 psia as minimum and maximum control points
- When storage tank pressure reaches 65-67 psig, maintain the storage tank at that level
- Turn on anti-ice purge flow

LH2 Fast Fill to 98%
- When LH2 ECO sensors, are covered, start fast fill by opening the transfer-line, topping and high point bleed valves
- Start recirculation pumps 25 minutes later
- At 85%, open the replenish valve, position the main fill valve to the reduced setting and close the inboard fill valve
- When either 98% sensor is 50% wet, or either 100% sensor is 15% wet, close the transfer line valve

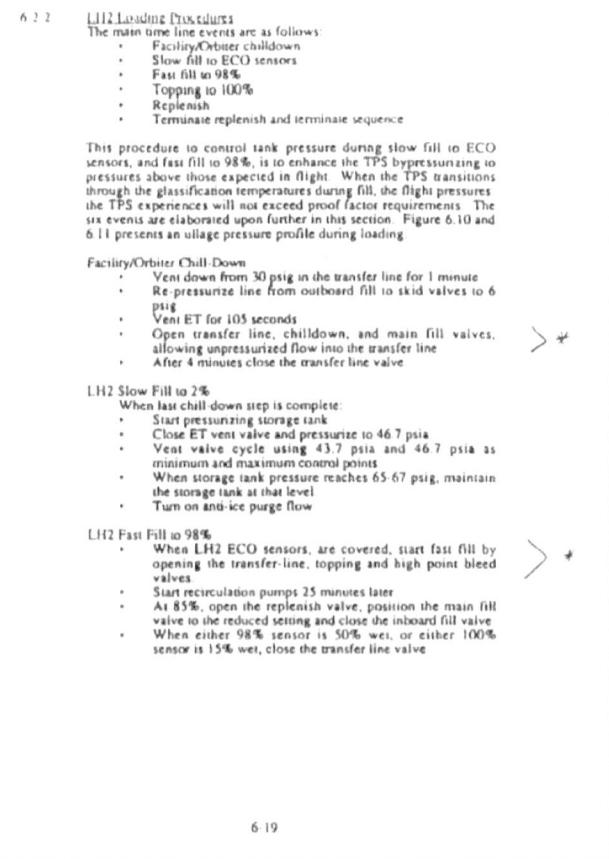

The Space Shuttle liquid hydrogen loading sequence. 1 of 2.

> L112 Topping to 100%
> - Continue to fill at the reduced rate through the chilldown, reduced main fill and replenish valves
> - Allow storage tank pressure to start decaying to 50-55 psig. Stay in this pressure range by cycling the vaporizer on an off
> - Open the vent valve
> - When any of the 100% or 100.3% liquid level sensors are 10% wet, start replenish by closing the main fill valve
>
> LH2 Replenish at 100%
> - Throttle replenish valve to maintain LH2 tank at 100% ± 0.2% flight mass
>
> Terminate Replenish and Terminal Sequence
> At T-1 minute 57 seconds:
> - Stop flow by closing the chilldown and replenish valves. Then open the chilldown valve to prevent pressure buildup in the transfer line
> - Seal off the engines and ET by closing the topping valve
> - Empty tail service mast (TSM) and Orbiter fill and drain line by opening TSM drain and purging them
> - Close ET vent
> - Turn off anti-icing
> - Pre-pressurize with helium to 41-44 psia
> - At T-19 seconds, close outboard fill valve
>
> As an alternative, the ET 7 procedure may be utilized. (see heavyweight Systems Definition Handbook).
>
> 6.3   Pressurization Performance
> This section details the pressurization line configuration, ullage pressure profiles for nominal and failure modes, and ET/Orbiter interface conditions. Data provided in Sections 6.3.2 and 6.3.3 is updated for each STS launch and incorporated as part of the Shuttle Operational Data Book update, prior to each launch. Data presented herein is based upon IVBC-2 loads and trajectories and baseline mission profile for the ET (reference the same document). The multiple failure conditions, provided in Section 6.3.3 are considered extremely remote. However, this data is required by Flight Control, as part of Mission Planning. Such data is used during flightcrew preparation to respond to such possible launch contingencies.

The Space Shuttle liquid hydrogen loading sequence. 2 of 2.

In 1988, an elementary introduction to hydrogen was among the first training classes I took at Kennedy Space Center. Hydrogen is serious stuff. It would stick in anyone's head that the spark of static electricity as you touch a doorknob was more than enough to ignite a leak of this odorless, invisible gas.

```
FLAMMABILITY AND DETONATION HAZARDS

THE PRINCIPAL HAZARD ASSOCIATED WITH GASEOUS OR LIQUID HYDROGEN
IS ITS EASE OF IGNITION OVER A WIDE FLAMMABILITY RANGE IN AIR
AND OXYGEN. DETONATIONS CAN OCCUR ALTHOUGH THEY ARE MORE
DIFFICULT TO INITIATE. SPARKS, FLAMES, DETONATIONS AND
TEMPERATURES IN EXCESS OF 1000 DEG. R WILL ALL IGNITE
HYDROGEN-AIR MIXTURES. THE COMBUSTION LIMITS OF HYDROGEN IN AIR
RANGE FROM 4% TO 74% HYDROGEN BY VOLUME. SUBSTITUTION OF OXYGEN
FOR AIR RAISES THE LIMIT TO 94%. THE LIMITING COMPOSITIONS FOR
DETONATION LIE BETWEEN THE UPPER AND LOWER COMBUSTION LIMITS.
FOR CONFINED OR UNCONFINED MIXTURES OF HYDROGEN AND AIR THE
RANGE IS APPROXIMATELY 18% TO 59% HYDROGEN BY VOLUME.

HYDROGEN-OXYGEN MIXTURES CAN BE IGNITED BY MINUTE IGNITION
SOURCES. FOR EXAMPLE, THE ELECTROSTATIC SPARK EXPERIENCED BY A
PERSON TOUCHING A DOOR KNOB IS 50 TO 1000 TIMES GREATER THAN THE
THRESHOLD ENERGY NECESSARY FOR IGNITION OF HYDROGEN-AIR
MIXTURES.
```

Introductory training material for the Space Shuttle liquid hydrogen system, 1988.

Hydrogen will get out, as the notoriously sneaky stuff reminded us when the Space Shuttle program ground to a halt in the summer of 1990. We thought, at first, a seal on the tank side (a 17-inch diameter "Naflex" seal) must be to blame. A tiny scratch on these beauties of aerospace engineering might quickly explain the hydrogen leak we saw while loading Columbia's external tank for mission STS-35.

With Columbia rolled back to the Vehicle Assembly Building, by Monday, July 2, 1990, we had dismantled the suspect line and removed the seal. Late in the day, some confusion ensued around who would inspect the removed seal. Would it go back to the manufacturer to examine for defects? Should it remain at Kennedy for us to analyze? In some quirk of events and confusion, I volunteered to transport the seal myself to our lab, carefully boxed, with stickers with stamps verifying who handled the seal when. Our first shift is over, just after 4pm, and I'm going down the road from the VAB to the logistics facility. The seal is in the back seat of my 1966 Ford Mustang. How do you drive a '66 Mustang with the seal in the back that could be to blame for bringing the Space Shuttle program to a halt? Also. Very. Slowly. Both hands on the wheel. Like our motto said, "Safety first."

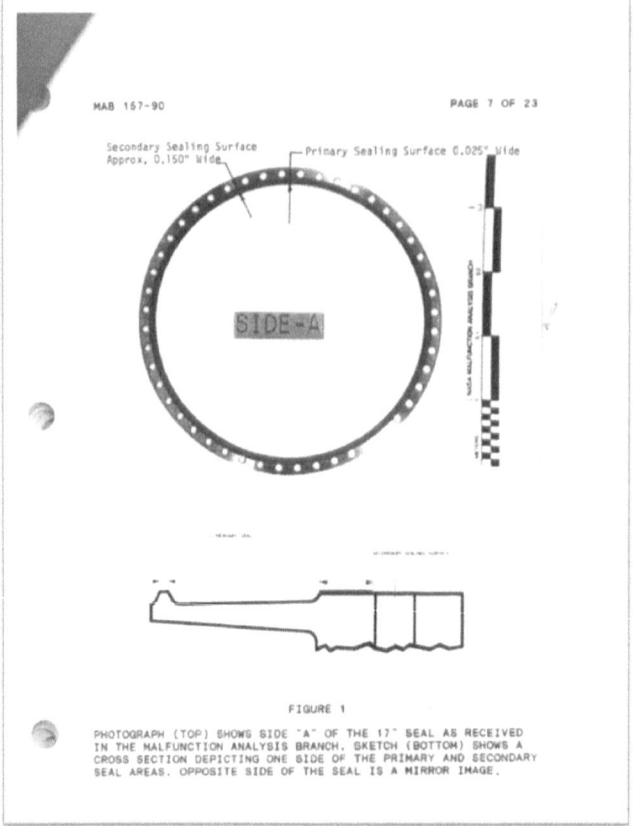

The external tank 17-inch Naflex seal from STS-35. Originally suspect, the cause of the hydrogen leaks in 1990 would lie elsewhere. NASA.

It turned out that large seal was not the leaker we were looking for. This would be a long, hot summer. The entire external tank umbilical assembly would eventually be removed to be tested in Rockwell facilities at Downey, California. A simulator for the orbiter side, we thought, would do the trick. The leak must be on the tank side, after all. (Placing the tank side 17-inch valve bottoms up would keep the hydrogen liquid in the assembly.) But being a sneaky leak, this testing marked merely a start. Hydrogen will do that, no other molecule being so small.

The External Tank-35 (also STS-35) 17-inch umbilical assembly, foamed over and mated to a test rig simulating the Shuttle orbiter-side at Rockwell, Downey CA, 1990. NASA.

Soon enough, hydrogen leaks plagued Atlantis too. At least the ensuing shuffle among flights gave us a rare moment when two Shuttles crossed paths in the night. This was somewhat of a fluke, as the original plan was to move Columbia and Atlantis on different days. Instead, the delays on one and the progress on the other worked toward a crossing just outside of the VAB, one coming, one going.

Atlantis, slated for mission STS-38, is parked in front of bay three of the Vehicle Assembly Building at NASA's Kennedy Space Center in Florida following its rollback from Pad 39A for repairs to the liquid hydrogen lines. Space shuttle Columbia (left), scheduled for mission STS-35, is rolled past space shuttle Atlantis on its way to Pad 39A. Image: NASA.

Our search for leaks continued at a test stand on Huntsville's Marshall Space Flight Center. This time we had the large umbilical assembly from Atlantis and the external tank side too. No simulator for the orbiter-side here – this was the whole contraption. The setup looked similar to the one we had in California, only now on steroids.

These were long nights, a particular day never seeming to end. This mid-morning, we knew testing would not begin until much later in the day. I was determined not to miss these tests as the representative from Kennedy, and the engineer who owned and operated these systems. This is when you go back to your hotel for a few hours of sleep after a morning planning session. Then you return at 2pm and stay till 5am the next day. (OSHA eventually came around to ask about this. The bosses signed waivers. Many waivers.)

Atop the MSFC test stand (left) sat the mated Atlantis and external tank umbilicals (right), ready for liquid hydrogen testing as we attempted to locate the leaks detected during loading at the launch pad.

Late one night, we were gearing up for more testing. As if hazards are not problematic enough at a distance, in a bunker with wavy foot-thick small windows, we debated when to send a "red-crew" up to the assembly as we flowed liquid hydrogen. The team members would use hand-held H2 detectors, "sniffers" we called them. Should the leak evade the instruments at the test stand, we hoped some hands-on testing would fill the gaps. Except just then, a storm started heading our way, intermittent lightning flashes switching surprisingly fast to continuous mode. I had some binoculars and was looking out a small window when I recall saying, "Maybe it's time to call it a night?"

> Wednesday August 29, 1990   17" Disconnect/Feedline Testing MSFC (ET37)
>
> [Page 20]
>
> 2235 hrs: Strip chart data verified & still coming in.
>
> ↳ Power out. Black out. All power down.
>
> 2243 hrs: Power back. Call it a day. Vent down. Secure.
>
> Note: Red Crew never sent out.

A lightning strike too close for comfort at just that moment left us in the dark. The universe answered my question. The sleep-deprived part of me liked the reply. Yes, it was time to call it a night. Clearly, that pesky leak had become our white whale, and our efforts now channeled Captain Ahab. We safely secured the hydrogen facility, with a good rest in order for everyone, amidst some mumbling about leadership not being mindful of people and safety. We would not be explaining an explosion that woke all of Huntsville and adjacent counties, leaving a crater where a huge NASA test stand should be. Or why we loaded hydrogen in a lightning storm.

By summer's end, we did find the source of those hydrogen leaks. Sort-of. As in not so conclusively and to everyone's satisfaction. But that is another story (and the very first paper I ever published, which I must scan and get out there.)

As NASA loads its new Space Launch System, a project now going on nearly 15 years, killed, lifeless, and then resuscitated at one point, the white whale is now in hand. Naturally, people will be watching the weather, as no one will be flowing hydrogen with a chance of nearby lightning. As we might tell a friend as they leave us for a trip, it's a good time to say to SLS, "make good choices!" and "be safe!" Also, watch that hydrogen. It just wants to get out.

And by the way, NASA, please share what's happening during the SLS test. One lesson of many I learned at NASA was sharing is a good thing. The world could especially use NASA's calm and reassuring voice just now.

* * *

Part 5

PART 5 COMMERCIAL SPACE – AND THE FUTURE OF NASA

Space, the commercial frontier. Somehow, it doesn't have quite the ring to it. Technology is fantastic, exciting, and sexy. Yet the phrase "commercial" gives off an odor of money and motives that are less than. The space business was slow to recognize the role of competition, incentives, and markets in bringing the benefits of space down to Earth. Were we an immature market? Or caught up in the national enterprise, the prestige, and the amazement of launches and pictures of planets? For a time, it would seem every industry except aerospace knew the difference between the gadget in your hand, effectively meeting a need or not, and the price, defined by the efficiency of putting it out there for anyone to buy. Fortunately, this is no longer the case. The conversation has shifted many times more to benefits, growth, finance, and risk, and new products coming soon. What is this "commercial space" thing really about, though? This is more than a new government modus operandi around contractual terms. It's more than a company trying to go it alone without NASA. It's the part about being *more* that will matter.

SUSTAINABILITY AND SPACE EXPLORATION

Oddly, one of the first books given to me when I arrived at NASA was for acronyms. Not what systems did or how they worked. Not flowrates. That would come later. First, acronyms. NASA had so many new things needing new words that it had turned grouping words together into an art. Somehow using only first letters sounded fluid. I now had a spell book gathering up all the incantations. "Confundo!"

There is a particularly important word a while now, "sustainability." My first recollection of the word becoming fashionable was around 2006. Every program suddenly wore the phrase. It actually took a while before anyone asked what sustainable meant. About then, NASA began its short-lived Constellation program (soon canceled as unsustainable).

Definitions of sustainable came and went. Some definitions nearly said a project with a strong stakeholder was the picture of sustainable. An ease to survive had become an ease to sustain. An inevitable budget meant easily sustained. Unfortunately, this confuses demand with supply and can even reward the confusion.

The United Nations defines sustainable as "development that meets the needs of the present without compromising the ability of future generations to meet their own needs." This is a much better definition than just some ability to go on, as it sets a bound on how to go on. This connects actions now to effects decades out. However, a good turn of a phrase may not do it. Sometimes it takes a good story, like "The Overstory" by Richard Powers. This wonderful book will leave anyone wondering if we understand the idea of what's "sustainable" at all. It's a story where trees are as central as people. You will never look at that tree in your yard the same way.

Unsaid in all this, what do we measure when asking what's sustainable. Do you look at the health of the forest or only check the orders for paper? Checking for a programs budget persisting is a lot like checking the orders for paper. We may see more orders, for a lunar rover here, a probe there, but we should be just as concerned with what capability is real and growing.

Even accepting this reorientation is necessary, it may not be sufficient.

There's an insight in "The Overstory" for our aerospace world. A sustainable space exploration program not only grows for future generations, it's healthy, diverse, and vibrant – an economic and technical ecosystem full of possibility. Repeating missions can be a lot like repeating trees, not improving the health of the forest. We can re-grow forests, copy of tree after tree, managed and curated, or as a character in Power's book reminds us, "You can also arrange Beethoven's ninth for solo kazoo."

Which leads to some thoughts on our space sector, first, the obvious, that being sustainable is not just about demand, and it's not just the next step either, curating some supply. It's more, about having a growing and complex system. Under every felled log, we should find a next level of connections, fine-tuned yet adaptable balances of competition, not only what's planned and predictable. In a space exploration economic ecosystem, we should too, or something is not quite right or sustainable.

IT'S NOT WHAT IT LOOKS LIKE – THE COST OF ISS PER YEAR

There is an oddity to the International Space Station, its name – a station. On Earth this would be fine, a station, as in stationary, not moving. In space, though, "station" is a bit of a misnomer for a facility going once around the Earth every 90 minutes and traveling 15,500 miles per hour. Pictures, even videos of spacecraft coming and going from the ISS, cannot do justice to its movement as it traces its orbital path. With that stationary bias, as if the ISS merely floats in space above us, it's also lost that the space station has a front. The forward Unity, Kibo and Columbus modules take the micrometeoroid hits along the way. Everything facing forward gets the dings. Unfortunately, adding a figurehead to the ISS, as on a sailing ship, to clearly show it is a ship traveling in Earth orbit, and where the front is, would be out of the question. An ISS figurehead at the bow would block the front door and the mating adapter where spacecraft dock to deliver cargo and crew.

A bow with a figurehead.

Similarly little appreciated is how NASA once upon a time had a plan to complete the ISS and soon after ditch it in the ocean. The ISS would have been built over decades to be operated for only five years. In a world of NASA projects that last a career, this qualifies as a blink of an eye. This massive and international project, the only reason to keep flying the Space Shuttle after the loss of Columbia, would be completed in 2010 and de-orbited in 2015. The flirtation with the idea of building the ISS and soon after destroying it was (fortunately) brief – a span of a few years, ending with acceptance of the absurdity of such a plan. I had the unique privilege in 2008 of briefing NASA leadership about the big picture for exploration, the ISS and everything else in the NASA budget, and declaring this end-of-life plan for ISS ridiculous (only in more diplomatic terms, with pretty graphs, and many budget scenarios and lots of numbers). As evenly as I was told I presented, this was a time when pointing out this apparent defect in planning was still a formula for an unexpectedly heated immune response (and not being invited back). We've been given our orders, which is what we plan around. Period. Until new orders.

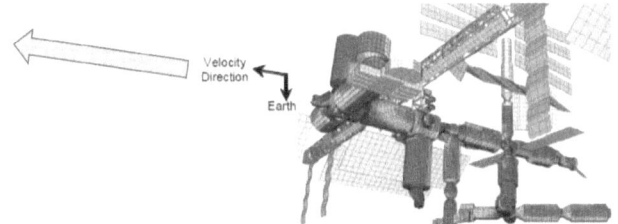

Fig. 2-3. Probability of impact from debris.
Note: Expect higher impact rate in "red" areas & have more capable shielding; "blue" areas expect lower impact rate & have less shielding

From Micrometeoroid Debris Shielding, Eric L. Christiansen, Johnson Space Center, NASA.

It's now 2021, and the ISS is still going strong. Current ISS planning looks as far as 2028, officially anyway, but practically, the ISS is part of NASA for the foreseeable future. In all this, a question repeatedly arises – what is the cost of the ISS per year? In 2010, the slight nuance to this question from on high was "What does it cost for NASA to *own* a space station," likely from a suspicion something was off about the other form of the question. The "owning" angle would not be better for questioning the station's yearly costs. Most answers to the ISS yearly cost question are "not even wrong" for many reasons. However, looking at the ISS books shows this also applies to the question. After years of being fortunate to be involved in uncomfortable questions, it became clear that asking the right questions is much more important than answers to poor questions.

Wrong questions, good questions and a trip down a rabbit hole

What might motivate questions about the ISS annual costs? A frequent, if not as advertised, reason is the belief that the question leads to a number that is the same as the amount of NASA funding freed up for other distinct projects should the ISS end. This is Deja-vu all over again, like the time over a decade ago when NASA planning called for building the ISS and then, not long after, hurling it into the Pacific. Today, the part about de-orbiting so soon is gone, but the notion that a NASA lunar program receives all the freed-up ISS funds for wholly new projects remains. This would help get those Moon plans to add up. Here again, the dull notion raises its head – that ISS funds, when ISS ends, are funds freed up to be used as down payments for a lunar lander, an outpost, or a giant rocket. And again, we have a question that is "not even wrong."

There is an excellent reason to ask some form of the question about the yearly cost of the ISS – a vision that someday, NASA will buy time and space for its astronauts on private space stations. Even better, NASA will be one of many power, air, food, water, ports, and space customers at those private stations. NASA would not own the lab, but it would rent the space like a room at the Hampton Inn. This is an entirely different reason for understanding the cost of the ISS per year. It is all about business cases that need to know where we are to understand what we must improve. NASA now runs with a business case for getting cargo to the ISS, the ride, not by using the entirely owned and operated Space Shuttle, but by hailing a cab. This became the US ISS Commercial Orbital Transportation Services program or "COTS." The business case then began with measurable NASA's needs - tonnage to the ISS per year, preferably at a lower yearly budget than continuing to use the Space Shuttle and its logistics module. Similarly, asking about ISS costs sets the stage for an excellent question. How might private space stations be had, where NASA rents rather than owns, for less yearly budget while also growing capabilities?

Entertaining a cost question about a NASA project gives new meaning to "going down a rabbit hole." In the ISS case, what will not be found down that hole is an answer equal to funding one day freed up for just anywhere else. There are three legs to this stool, most of which, by dollar amount, do not have to do with operating the ISS.

The first cost is basic "research" about humans in space, a body of knowledge unlikely to be declared complete anytime soon. This is what is done on the ISS, not the cost of running it. In the 2021 NASA budget, this is about $350M a year.

A second cost usually attributed to ISS is the cost of getting there. When asked about running a lab, the standard answer is to add the flight cost, supplies, and the rental car to get there. This is $1.7 billion a year in 2021.

Third is the "ISS Systems Operations and Maintenance," the actual cost to run the station at about $1 billion annually. This is how anyone googling ends up with a reply to the "cost of ISS per year" of "$3 to $4 billion a year" (also here). This is the sum of these three parts. It's in diving into these three parts that it's easy to realize the confusion afoot.

Google's answer: Nyet. Not quite the answer. Not even close.

Going further down the rabbit hole Alice?

Dissecting is a good word for looking deeper into ISS costs, as this will get messy. For one, NASA will unlikely cease its R&D into humans in space. If NASA moved to a private station, the thought might be equivalent research could be done for less, but just as well it might invest more by choice. NASA could finally dedicate more resources to results and less to the means of getting these. Tossing these comparisons into spreadsheet territory would distinguish NASA astronauts from others (international partners) and then only do research, minus the effort to keep things humming along. The private space station equivalent research is not seven astronauts. But is it two, five, or in between? In cost circles, when confronted by valid but competing reasons why a cost might go up or down a first good guess is it will probably not change much at all.

Similarly, doing the research needs astronauts, and NASA currently manages to get its cargo and crew to the ISS commercially. A private station might do better if also tasked with the ride, now a shuttle bus to the hotel. Still, it could arguably be unlikely to do much better given that NASA has already acquires these services in a very competitive field (Northrop Grumman and SpaceX, with others to come). Current post-ISS thinking already recognizes this - the transport to the ISS is being commercialized, and transport to a private station would not be re-commercialized. Any changes here will result from how costs are attributed to non-NASA crew and cargo - employees of the private station. (As well, showing more morphing than transformation, future transport of cargo and crew would remain funds for transport of cargo and crew, just to the Moon.)

After the research and the transportation to a space station, the last element, which seems to be the ISS yearly operational cost ($1 billion a year), requires a secret decoder ring that comes from years of experience asking where the bodies are buried. However, anyone who dares go down that rabbit hole will discover that even that question is not valid. The better question is *how are the bodies buried, where, and why.* The number here gets clearer when realizing that NASA budgets for space exploration have always included Kennedy Space Center's expected and necessary ground and launch operations for when the day comes to prepare and launch deep space missions. Yet, the budget has been missing when looking at these same budgets for the expected and necessary mission and flight operations at Johnson Space Center.

This is part artifice and part policy. As long as NASA is in the crewed space exploration business, many of these human spaceflight costs are attributed to whoever the current customers are – or now the remaining customer - the ISS. When the Space Shuttle ended, a look at the NASA budget clearly shows this institutional effect. The Space Flight Support budget shot up, and many people and organizations from Shuttle moved there. These people (and the dollars) were *not freed up for new hardware with the Shittle ending.* (The similar effect here for when the ISS ends significantly affects the possible budget advantage should NASA move its research to a commercial station.)

Are you confused yet? Searching for better questions.

If all this seems to say, the operation of the ISS is nearly free to NASA, apart from getting there and actually using it to do research, it's not. It does mean the commonly googled number of about $3 to $4 billion annually for ISS costs is misleading. This is a gross error if you think this ISS yearly dollar number equals the dollars usable for other project possibilities when the ISS ends.

Yet a poor question has its uses – it can lead to better questions.

What might NASA do in a permanent place (or places) in LEO as an anchor tenant, renting a room or rooms but not owning the hotel and lab or part of an international one? Aside from doing the numbers on a business case for NASA (and another one for the private partner finding non-NASA customers), might we see NASA do even more in orbit for the same yearly ISS funding? If doing more, what more? With byzantine accounting, the benefits we should focus on are too easily lost in the fray. *Rather than abandoning our rear on our way to deep space, the question is how to strengthen it so that deep space voyages are possible.* If we are to have orbitals one day, a long string of pearls from Earth to the sky above and beyond, these are the right questions.

COMMERCIAL SPACE STATIONS BEGIN SHIFTING THE CONVERSATION TO "WHY SPACE"?

The familiar refrain "it's impossible to keep up with so much happening" has come to the space sector. Though this could be said in all walks of life. As we join the club, it's a good time to ask "why space"? Our aerospace industry is not unique, carried along in a wave, wondering if there is a rhyme and a reason.

HOW

As an engineer, I found myself in the company of other engineers day in, day out. Surprisingly, asking "why" about much anything was not the most popular question. If Yoda were an engineer, he would say, *"How, or how not. There is no why."* This is our mission as engineers - to figure out how to get it done. "It" being someone's requirements. In this world, figuring out how to meet a customer's requirement may be the only question that matters.

We had invited some new faces to the kickoff meeting. That's NASA lingo for the first big meeting, but not the first-first meeting. The project was half architecture and half technology. So, we needed those rare characters who intuited the laws of physics, knew their way around the labs and could work with the imagineers on how the pieces might fit together. Realizing we had been stood up by one of the new team members, we dropped by to inquire. Calendar glitch, life, the kids, a last-minute conflict? It was none of these. "I just want to do engineering. When you have your requirements, I'll help design the solution." In English, we'd just been lectured–"Come back when you know what you want."

WHO (and sometimes WHEN)

If you are higher in the food chain, it turns out how to get astronauts to orbit, or a probe to Jupiter or land a car-sized robot on Mars is not such a big deal. Instead, NASA project budgets depend on Congress and the Whitehouse. If this process were the eye of Sauron, it never turns its gaze too far from "who." Once upon a time, "when" was also important, so it was possible to create a space program and put men on the Moon in under a decade. National pride and clearly showing the superiority of a way of life was too important to leave to just someday. But in a world where budgets are tamer, it's now fine if when means whenever.

Image: NASA.

The planet Askiruh, or as close as the pronunciation comes, missing other guttural sounds, and a nuanced color display as the natives say it, was larger than Earth. Like Earth, it was small enough to have lost its light gases, but unlike Earth, it was larger enough that getting anything to orbit was practically impossible. When Askiruh technology was finally up to the task, a launch vehicle larger than any ever built on Earth was constructed to great fanfare. The tiny satellite, about the size of a Ronafor, a native rodent-like creature, barely made it to low orbit. For Askiri, when the film returned, seeing their world for the first time, as a whole, was a defining moment. The task done though, driven mainly by rivalry among certain Askiri, further efforts languished. Reasons for again expending significant resources for orbital dreams have waned in council debate, though limited sub-orbital space exploration continues.

"Orbital is hard." -Askiri engineers saying.

It's no wonder that getting to orbit and staying in orbit, with roots in how and who, persist and satisfy as ends in themselves. In most any industry though, transportation and facilities are a means to an end, not an end in themselves. That is unless you are a transportation company, a FedEx, or the JetBlue, or the owner of the facility promising certain amenities. Everywhere else, you worry when transportation is more than about a fifth of your costs. And your facility costs should be even less.

Yet in NASA, one glance of the Human Spaceflight budget reveals that nearly all its resources go toward getting to space, followed by the place to stay. Only the smallest part of the NASA Human Spaceflight budget does what might be called "production" once you are in space. These are a smattering of budget lines, like a few hundred million for research and development on the ISS. That's out of an over $10 billion-dollar Human Spaceflight budget, or 3 percent.

AN ENGINEER'S JOURNEY IN NASA

Image: NASA.

WHERE (as destination)

The presentation about getting to Mars was the customary 10 pounds of information jammed into a quart Ziplock bag. Then followed the 15 minutes, no wait, 3 minutes, no wait, "just time for a couple of questions." The first question was more a statement, proving the audience member also knew a way to get to Mars. The second question was about costs, so unwelcome. "We have plenty of resources. They're up there, right now, just going around in circles. The ISS." There was a moan or two in the audience (the questioner included.) Somehow going in a straight line to Mars was just better than going around in circles, and no reason was necessary.

As if getting to Earth orbit and staying there are of late resolved, NASA is now focused on returning to the Moon. It's easy to consider exploring a destination to be an end in itself, producing scientific knowledge. This would be knowledge beyond getting there, or operating facilities there (or thereabout, in lunar orbit). It's easy to allow a means to an end, the transport, or the facilities, mostly how and where, to again captivate the headlines.

In the same vein, of late, we have sub-orbital and orbital tourism, with the usual fanfare about how these may lead to cheap fares for everyone else. The analogies to air travel, once the domain of the barnstormers and then the wealthy traveler, must strain themselves to explain current events. Climbing

Mt. Everest as a discussion about risk, individual achievement, or inspiring others soon enough runs out of oxygen, running to still other analogies.

Finally, we are forced to come full circle to "why space?" Commercial space stations may be the first and best chance to shift the conversation. A trip and a place are merely the start of the talk, not the end as if no further questions are required.

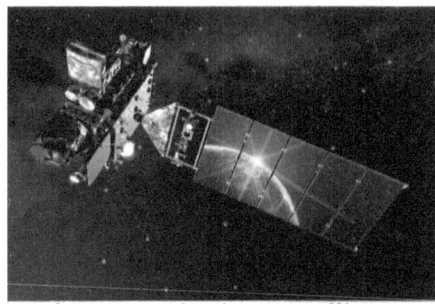

Weather satellites and communications satellites provided the earliest widespread benefits of getting technology into space. Image: NASA.

Images of human insulin crystals grown in 1-g (left) and microgravity (right). Crystals grown in microgravity are larger and of higher resolution. Image: NASA.

...And WHY (finally)

The small-batch run was invaluable. Like the starter for bread, it would now go back to Earth, where the company would cook endless copies. This was a monoclonal antibody, and not just any protein either. It was a crystal perfect for imaging as well. This would save lives, targeting cancers with the precision that made prior therapies seem like random shots in the dark. The impressive commercial space station would leave anyone wondering how such a concoction could be affordable at all until they realized the facility cranked out hundreds of these small batches for all manner of disease. And that was just half of the space station.

We could see this conversation shift sooner rather than later. Inverting the pyramid, most of the energy and effort (and cost) will no longer be the ride or the stay. Instead, most resources will revolve around results once you are up there. If Mars is always 20 years away, perhaps so too this moment, until now. We may soon look back and see the answer to "why space" was clearer as the focus of it all became less about the trip and the stay. Space is hard. Developing new products that benefit everyone's life on Earth will be even harder. But it will, like communications and weather satellites, again answer such a good question – "why."

NEW SPACE, A RORSCHACH TEST

Depending on the news, "new space" is commercial, innovative, well-funded by billionaires, and changing the world. The site of a Falcon 9 booster returning to land after being flown eight times tells a story of change, a revolution that, as predicted, is being televised, now in high definition. Crews that are not NASA astronauts have launched to the edge of space in Virgin Galactic's VSS Unity. Soon more people may go sub-orbital on Blue Origin's New Shepard or into orbit aboard a SpaceX Dragon. This is not your father's astronaut. And that's a good thing. But right behind, we hear the other story, how "space is hard," with news of an Electron rocket failing again, not far from a previous failure. Or we hear the new ULA Vulcan launcher engines are not ready as planned, indicating the heavy lift, reusable Blue Origin launcher will (again) be delayed as well. Yet it's easy enough to pause a moment and get some situational awareness, even if we know we must keep moving north down the valley and onward up the mountain.

We need a daisy-field effect, where an Electron or a Falcon is one daisy, and then you zoom back to see the field full of flowers. Sometimes, we need to step back, up and above. I have the occasional picture of the Space Shuttle taken at an odd angle. These are my favorites, taken from somewhere people don't usually crawl, a reminder there is more to what is under heaven and Earth than you see in the stock footage. Global commercial launches could use a similar odd angle. For a starting point, we go to the US Department of Transportation.

A unique view when crawling around in-between the Space Shuttle External Tank and the Shuttle Orbiter Discovery, at the launch pad, 2005.

While something is to be said for "minimally processed", the DOT data gets at least some crawling through and around for a better look. The DOT does not include the Vega rocket or the recently launched Virgin Orbit 747 LauncherOne rocket, so add those. A review to make sure everything is in order shows some discrepancies, like the DOT including the Dragon demonstration flight but not the Cygnus doing the same. Include both to be consistent. Do a dozen similar puts and takes of a launch here or there. Add launches since 2017, and you will have a curious picture of global commercial launches. (Left pending is completing the picture for launches from China. These might not be entirely commercial, but for all practical purposes, they will compete for some part of the global orbital launch business.)

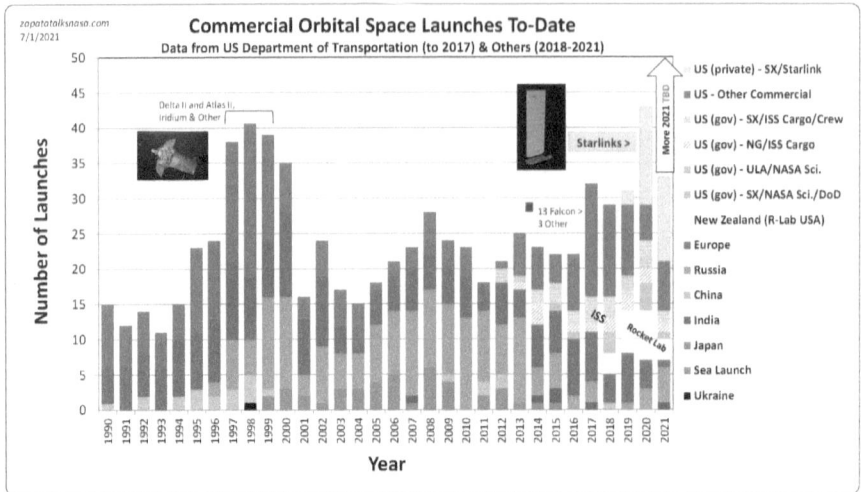

Global commercial launches since 1990 up to the Soyuz-2 launch of OneWeb satellites on July 1, 2021.

As the picture develops, there comes a sense of déjà vu if you've been here before. In the late 1990s, there were also a dizzying number of launches, such as the Atlas and Delta rockets launched for the private sector, carrying the first dreams of low Earth orbit satellite constellations. There were 41 commercial launches in 1998 compared to 43 in 2020. The "LEO Launch Frenzy" was driven by the dream of satellite phones. The Iridium constellation, in particular, was supposed to revolutionize global communication and change everyone's life as we knew it – the story is that the name came from the layer of Iridium that marks the death by asteroid of the dinosaurs. As happens, a cell phone in every pocket would change the world and us, except with cell towers here on Earth, not in orbit.

Oddly, Iridium lives on to this day. I could not imagine that in 2017, I would rent an Iridium phone. Hurricane Maria had wiped out the power grid in Puerto Rico and all communications, from landlines to cell towers. Iridium had placed enough satellites in orbit years earlier, and when bankruptcy came, they had a functioning and somewhat capable network in the sky. It operates to this day. It's curious how an ambitious business case can close if you don't have to pay back your up-front costs (and creditors). One day, you somehow have that oh-so-important connection that only needs a handset. Arriving in Puerto Rico a few weeks after Maria – dinner with family by lamplight after the adventure of getting there – priceless. Knowing there was a means of communication if we needed it – not bad.

Which, of course, brings us to the 800-pound elephant in the data - Starlink. Oddly, Starlink is excluded in some data sets about commercial launches or satellites because one company is not paying another to make or launch satellites. I have instead decided to acknowledge the gorilla. More so, I've given the gorilla its own seat at the table by separating Starlink launches from other commercial launches. This is not entirely arbitrary. The DOT definition of what is "commercial" states that a commercial launch is a "launch that is internationally competed (i.e., available in principle to international launch providers) or whose primary payload is commercial in nature." It would be a stretch to say Starlink is not "commercial in nature" and exclude it when its reason for existence is to revolutionize commercial broadband.

Here, we begin to see differences between the late 1990s and today in quality, if not quantity. For one, the technology of the Starlink satellites is a generation removed from that of the 1990s. The form factor recalls a cell phone, flat, compact, ready to connect and spew news, gossip, and tweets. It also recalls the skateboard platform of a Tesla and now most electric vehicles. Laser crosslinks between Starlink satellites can eventually make sure that if one satellite cannot find a ground station, it will simply find another satellite that can. For those who show their age, like us vampires reminiscing about summer in Paris in 1892, we know what we called a computer in the heyday of Iridium in 1998. Then we see what we call a computer today and appreciate the difference.

More so, seemingly forgotten in the frenzy of Atlas and Delta launches in the 1990s, there was an expectation of a low-cost launch then - "coming soon to a theater near you." Except that movie never arrived. At the same time LEO constellations were trending (before trending was a thing), all kinds of launch vehicles were also in the news (if not in the air). Reusable launch vehicles were uniquely linked to the eventual success of LEO constellations like Iridium, Teledesics, or Astrolink. Every constellation seemed to have a launcher, a **Rocketplane**, a K1, a **Roton**, or an **Astroliner**. If a constellation needs lots of launches quickly, a reusable launch vehicle needs lots of launches to close its business case. The twain would never meet.

And so, another difference since the 90s. There actually is a vehicle today that has much lower costs. As foreseen, it's also reusable (mostly). There is no waiting today. The business case of a satellite constellation is no longer at the mercy of someone else's business case for a launch vehicle. That vehicle with lower costs – the keyword being costs – actually accompanies the Starlink constellation. For SpaceX's Starlink, a Falcon 9 launch, already affordable in price, is likely even better - an internal item on the ledger at "cost," not "priced," with a profit and all that on top.

That's only the start of the Rorschach test on the global commercial launch chart. Do you see butterflies or tigers? The seeming symmetry between 1998 and 2020 is very likely superficial. The facts on the ground (and now in space) are that advances in satellite and launch technology could make all the difference between failure then and success now. Still, a question was explored back then that could prove just as important now. Market elasticity.

If we build it, will they come? This has been the eternal question in space launch. An "elastic" market sees more sales when prices drop. More so, the launch supply grows, and new products emerge with the demand for the rocket. We cannot imagine what will go atop lower-priced rockets one day. Growth makes the pie bigger, instead of everyone trying to grab more slices of a small pie.

Some work trying to figure out this question, like the 1994 Commercial Space Transportation Study, cheerily concluded there was much promise. Launch market elasticity was there, if not in every market, then in just enough. Demand would increase as launch systems reduce costs and pass those improvements to customers as reduced prices. The ideas for what goes atop the rockets would flourish, like Starlink, but also rides for tourists or private space stations if you want to get some work done or just to take in the views. A look at global commercial launch data would seem to be the way to answer the question coming from the back seat – are we there yet?

Like an economist, but not the one Truman asked for, on the one hand, it's difficult to imagine all those Starlink launches without lower Falcon 9 costs. In this sense, the drop in Falcon 9 "costs" caused (internal) demand to increase, deciding to build a Starlink constellation, the very definition of market elasticity. On the other hand, that lower internal cost is not a "price" or the whole market. A functioning, healthy market has many players, different buyers and sellers, and never just one buyer (like NASA) or seller (like SpaceX.)

Still, there are promising signs of growth. Since 2017, the uptick in global commercial launches has held firm. It is not quite enough yet to draw a strong trend line, but it is definitely enough to give pause. The knee of an S-curve looks like a little squiggly blip just after nearly nothing but a faint heartbeat. Yet many a knee in an S-curve has resulted in a revolution or two. One day in 1983, we got a kick from seeing the "car phone" the company gave our brother. The battery weighed about as much as a small lead acid battery. (Oh wait, it was a small lead acid battery.) In the blink of an eye, it's 2000, nearly everyone has a cell phone, the thing fits in your pocket, and companies that didn't pay attention are in bankruptcy. Of course, we still complain about the battery life.

The uptick in global commercial launches since 2017 has as much to do with small launches and the Rocket Labs

Electron as with medium launches and Falcon. All this would seem to have come at the expense of the Russians and the Proton, but this year has seen a little more life in that Proton, enough to not write it off - just yet. The end of the shuttle and the start of the NASA commercial cargo and crew launches have also helped the uptick since 2017, with a few launches a year, give or take, since 2015. NASA is now part of this equation of commercial launch, not just as had been the case with the one or two science launches a year procured on a commercial basis. NASA's "commerce" is part of an international supply chain that keeps the International Space Station (ISS) safe, sound, and robustly stocked.

To be elastic then or not to be elastic?

It would seem the data is still being coy, hiding the answer. Amazon appears to think "prices" are already low enough, buying nine Atlas V launches to get started on its Kuiper satellite constellation to compete with Starlink. However, if you have a mind to launch most of your constellation on mostly reusable Blue Origin New Glenn rockets, that will also be an internal cost, and the real deciding factor may just be a rush sparing no expense - for now. In this view, the curious question stops being market elasticity for launchers and becomes market elasticity for launchers/constellations as a whole system. Just when you think you have some answers, the question changes.

Lesser known, there was a LEO constellation planned once upon a time by Boeing, a brief flirtation with manufacturing and operating a system. This would have been quite the departure from Boeing as a manufacturer, handing off hardware to customers, the airliner to airline model. The DARPA XSP, my last project before retiring from NASA, was linked to such a constellation. Once again, there was a custom-made demand to justify the investment in XSP to supply improved transportation. Boeing opted out at the end, on its satellite constellation first, and not surprisingly, on the XSP later. Why get all dressed up with no need to go?

Perhaps the Rorschach chart should have launches besides operational satellite count or bandwidth deployed? It could as well try to show a measure of faith, as much required as supply and demand. A business case is one thing, an ability and a desire to take risks is another. Not too long ago, at a "big aerospace" meeting, our lead segued into a discussion about risk. The question posed to the room full of industry veterans and some just starting off was simple – a long time ago, your company took some significant risks that could have sunk the massive company you are a part of today. Would you see doing this again? We had been around long enough to know this was a lawyerly question, one we knew the answer to. The purpose was to move the discussion toward the topic of risk. Enough of an answer would have been some blurbs about being a company with responsibilities to shareholders paying predictable dividends to people who like predictable. Surprisingly, the room went into a cacophony of responses as everyone responded at the same time. It was a good question.

At its current rate, there will be 47 Falcon 9 launches this year. It's a simple calculation: the launches to date in 2021, how often they launch, and how many you get if the pace keeps up. It's also unlikely, but still, at any rate, this year, some single SpaceX rockets will be reused more often than the entire launch rate of other launchers. To harken back to recent years, we saw about 40 commercial launches globally. Now, we have a launcher that is at least capable of doing that on its own, with a reusable fleet of boosters.

We will always need unique angles, crawling around in data to get a picture that says more than the traditional shot posed and at a distance. We are awash in data, investments, public and private, satellites deployed, space debris, and valuations. In the end, what will we see in these pictures? Today, I see growth and possibilities that did not exist before. I also see that by all indications, this time is different.

ARE YOU HAPPY ON AVERAGE?

It was only some years ago I wandered upon the word "wonk" or "wonkish" as a reference to someone diving too much into obscure details. The definition implies annoyance with too much detail. First come the numbers, then come the graphs. The predictable debate comes right along – does X cause Y, or is it the other way around? Maybe there is no causal relationship at all. If you got that far, you too might be wonkish.

Oddly, to most analysts, like me, one thing graphed against another, often just the year, is what we call the big picture. It's our view at 100,000 feet. To the audience, this seems excessive detail. Go figure (or graph, as you prefer.)

There is no lack of graphs for launch costs, or really prices a customer can expect to pay, the cost to them. It's safe to assume NASA or any US customer's launch costs are affected by the purchases of others, even if we don't have access to the books. A company might have a certain customer pick up certain costs for example, rather than spread them among all customers. Why not assign a little more here, a little less there? For example, NASA launches crew and cargo to the ISS on Falcon 9's and employs the United Launch Alliance Atlas V for its uncrewed scientific missions. (And eventually, an Atlas V will be used for NASA crew with Boeing's Starliner.) So perhaps it's worth getting a bit wonkish here and seeing what the data says about the costs of launch to the US Department of Defense.

Cost data for the National Security Space Launch program is published in a valuable report from December 2019 entitled "Selected Acquisition Reports." This report is a high-level summary, to an analyst anyway. It shows funds and such across time, helping with a couple of questions – *Has the cost of launch to the US DOD dropped in recent years? Where is the SpaceX effect?*

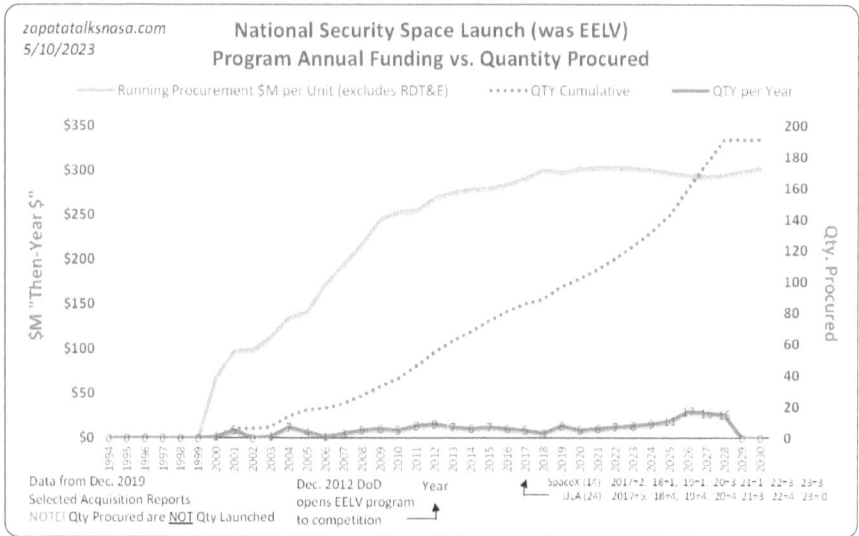

Graph of National Security Space launch funds per year using data from the December 2019 Selected Acquisition Reports. (Disregard the oddity of turning up sharply then flattening at the end. To the US DOD, a program ending is like approaching a black hole, where the laws of physics as we know them, and common sense, break down.)

To get our bearings, the US DOD first introduced competition into national security space launch in 2012, but the first SpaceX launch for the DOD would not be till 2017. It was easy to predict then - *change takes time.* 2012 was also when US government auditors reported massive cost increases in the then-EELV program. As well, 2012 saw SpaceX dock its Dragon cargo spacecraft to the ISS for the first time. Still, it would be five more years before SpaceX launched its first secret payload for the US National Reconnaissance Office.

Now, more years have passed, so we have more data. Gather up enough numbers and who knows, we might see a line form that says something. For example, the data shows that 2012 was a peak funding year, so the DOD buying launches had a peak "cost." Since then, the good news, annual budgets for the Space Force buying launches have steered well clear of that peak $3 billion mark.

Of course, the question that follows is about costs per unit, the real indicator of the US DOD seeing savings or not. Here, there seems to be no significant drop in costs on a per-unit basis, at least in then-year dollars. There is no nose-dive in the line, that is. Yet a price holding steady over time, assuming there is inflation, is a price dropping in real terms to a customer. Notably, it seems 2012 was a turning point, after all. US DOD launch costs (per unit) climb steadily up to here, then level off.

There is no more recent SAR report, but we know the yearly budgets here for 2020-2022 are significantly less than the numbers predicted in late 2019. Then again, there has been some drop-off in the launch rate, so the jury is still out. Officially, the DOD maintains it has saved billions since 2012. It would seem the data somewhat supports this – especially judging from a sense the running procurement cost per launch could have stayed on its original (unsustainable) trajectory.

What works against the elegance of all this X leads to Y and wonking out is those pesky SpaceX launches. SpaceX Falcon 9 launches for the US DOD have typically scored in the $90M range, leaving a gaping question. If SpaceX is the lesser launch provider by quantity, where is the rest of the yearly funding? An attempt to put an average running cost per unit is likely having one foot in boiling water and another in a bucket of ice. Is it fair to say that we are comfortable with the new temperature *on average? Does the question have the meaning we started with at all?*

All this could be settled. Some years ago, the US DOD kept abundant data online *and public* in "RDT&E" reports. These were yearly, very detailed, if excessively so, and helpful in ways a summary SAR report is not. (The detailed reports feed the summary one.) These and other public access sites have all slowly gone dark, ironically, just as the competition heated up between SpaceX and United Launch Alliance. Of course, it's all only a wonkish interest, as launch costs dropping for something as important as US national security space launch may be just an annoying amount of detail.

FINANCIAL RISKS, SPACEFLIGHT, AND THE QUESTIONS WE ASK

"Would Boeing make a bet like that again, on a low-cost space launch vehicle," asked our team leader. Suddenly, thirty or so people burst into a Tower of Babel, everyone talking at once as if a spark set off a conflagration. Mostly the cacophony of replies leaned toward "no," or jumped right into statements - how different our world is today versus then, how we now have shareholders, or how the technology is not the same. Unlike in the movies, this moment in the meeting did not slow dance a conversation. There was no pause by someone at the table, followed by another insightful contribution from the next character. I saw again that everyone talking at once is not atypical for engineers and scientists, that is when faced with a question about risk.

Starliner approaching the International Space Station. NASA

Boeing "bet the company," in aerospace lore, when it built its first jet airplane in the fifties. Now, in 2016 we were visiting Boeing to see what thoughts their engineers had on low-cost access to space. The hosts and the visitors knew each other well from many years and projects. Someone said we had gathered all *the usual suspects*. Fast-forward to this week, and we see the Boeing Starliner crew spacecraft arrive at the International Space Station, minus crew for this demonstration flight. This milestone is part of a tectonic shift in how NASA relates to its private-sector partners. Boeing signed up to build Starliner for a firm-fixed-price rather than the usual cost-plus, by the hour contracting, with its uncertainty about the total cost.

This is not the first time a firm-fixed-price contract has been in the news, as these are increasingly common for NASA, and are even appearing in the US Department of Defense. As NASA spaceflight sees the advantage of partnerships focused on results, such contracts are now more the norm than the exception.

Financial risk for a partner is usually not emphasized in these new arrangements with NASA, perhaps because most have been so successful. Nonetheless, Boeing's Starliner is a poster child to remind us about financial risks, as with other projects Boeing has signed up for on the dotted line. To date, the Boeing Air Force One deal is a loss of a billion dollars for Boeing. The KC-46 tanker aircraft for the Air Force, also a fixed-price development program, stands at $5.4 billion in losses.

That's more in financial losses for Boeing than the Air Force will pay in total. Tankers for sale, 50% off! Boeing's Starliner program also "booked charges" in recent years, accounting-speak for costing that much more than you got paid. These losses were $410 million in 2020, then another $185 million in 2021.

With a little (ok, a lot) of forensic accounting, it's not difficult to draw out some curiosities beyond the big numbers. The NASA Commercial Crew program awarded a $4.2 billion firm fixed price contract to Boeing in 2015. Later (with much criticism), NASA gave Boeing an additional $287 million. So, it can be said we have two numbers for a Starliner program: there is the Starliner cost to NASA, $4,487 million, and there is the cost to Boeing, $5,082 million.

But those totals get ahead of things, as what are the total dollars for? Here, NASA mixed apples and oranges, development and operations, to give its partners more latitude in the initial commercial crew awards. To further complicate matters, quantity varies too. This is not just apples and oranges, it's different numbers of each. (The critical contract line is *"The minimum quantity of missions to be ordered is two and the maximum potential quantity of missions which may be ordered is six."*) Operational flights come after development is complete, as marked by a first successful crew demo flight. When all is said and done, seemingly, SpaceX will have delivered 6 operational missions before beginning its subsequent contract at the Crew-7 mission (for a grand total of about $2,600M). Boeing will have delivered only two operational missions (the minimum) if it goes into its next contract at the Crew-3 mission (for a grand total of $4,487M). Alternately, an interpretation might conclude there are more "charges" to follow if Boeing is to deliver the promised launches ahead, with a crew.

It's tempting to say an answer to our question from 2016 is now in much better shape, from experience rather than from a survey of the room. Would a company, around a long while or just born, bet its future on knowing NASA investment is just a start, while new opportunities created by lower costs (safely) are the end? Commercial partnerships have been an enormous success for NASA since their beginning. In practice, the private sector continues to sign on when asked to take the bet. This is so even in the face of risk, with technical difficulties becoming financial losses.

After Starliner will follow a Dreamchaser for cargo, perhaps one day with crew. Private spaceflight has also taken off with personal Dragon flights, Inspiration 4 just the start. Maybe just as the relationship between NASA and the private sector changes, it's time to change our question. When NASA takes a risk and asks if others will join, can any company afford not to say yes?

NASA COMMERCIAL SPACE, THE 16%

A bit wonkish, again.

If NASA commercial space is a thing, how much of a thing is it? Numbers can help provide part of an answer, but not all of the story. For that, we need context about NASA's commercial programs, the rest of NASA, and the world in which NASA lives. Also, there is the 3rd law, where every action has a reaction. Having shifted with the times, is the shift likely to gain momentum or create pushback?

By the numbers, in 2022, about 16% of NASA dollars will be spent on its commercial programs. A commercial program typically involves the government contracting with more than one provider focusing on a milestone, a result, or a service, at a firm fixed price. What else is there? Wouldn't anyone spending this much money always have lined up some competitors to assure there is a backup? Wouldn't the price always have been agreed to up-front? Wouldn't the payment always have been withheld until delivery? No, no, and no. On the contrary, such commercial arrangements were rare for NASA until taking this route for getting cargo to the International Space Station, knowing the Space Shuttle's days would soon be over.

This innovation by desperation, it's been said, decided to

take a risk and see what industry might do for much less cost if NASA were much less bureaucratic. It turns out you get much more for fewer dollars when it's clear no more dollars are on the way, *and* you let industry innovate their way around this sea-change. Armies of analysts, number crunchers, and policy wonks are still trying to figure this out. Fortunately, most do not charge to these new projects.

For a bit of comparison, NASA's overhead costs in 2022 stand at about 15%. That is, NASA is spending as much money on its new commercial partnerships with the private sector as it spends on overhead. (This is from taking NASA's 2022 "Cross Agency Support," a hodgepodge from facilities, to I/T, to physical security, and more, versus the yearly budget.)

Viewed over time, though, that laundry list of NASA's overhead (or "support") costs across the country and its projects was 21% ten years ago. Not that anyone wanted to step up and say that NASA needed to spend more on support. Naturally, overhead was told to make do with the same money year after year, dropping as a percentage as the NASA budget increased. The opposite happened for NASA programs opting to go commercial, which stood at only 9% of NASA spending ten years ago. While it's tempting to wonder if NASA shifted overhead into commercial space hardware, the actual flow of money is likely very different. Albeit a dollar being a dollar, anyone pondering the fungibility of it all is asking a good question.

Ten years ago, too, NASA's commercial cargo program was well underway, with the first SpaceX Dragon cargo vehicle reaching the ISS in May of 2012. The commercial crew program followed. More recently, we see the commercial approach in a lunar lander and a lunar Gateway. If you plan to land on the Moon, you need a lunar lander. If your trip planning to get everyone to meet up includes a gathering in lunar orbit, you'll need a small space station there too. Portions of these are commercial, fixed price, the power and propulsion for the Gateway, for example. Other portions, like the Gateway crew quarters, may have been rather traditional contracting at first, ending up firm fixed price.

Jumping to 100,000 feet to take in the view from above, we have four major NASA commercial space programs. Two of these are operational capabilities for getting cargo and crew to the International Space Station. The two other commercial programs are in development, the lunar lander and the Gateway, to eventually put NASA astronauts on the Moon.

To be complete, though, NASA has a fifth major, operational commercial space program that gets little fanfare. It began way before commercial space was a thing. The NASA Launch Services Program is a NASA commercial space program of little notoriety. Here NASA buys launches on a commercial basis for its robotic science missions. Formed in the early 1980s, this program is located at Kennedy Space Center, which is also the home of the NASA commercial crew program (Florida).

Interestingly, the commercial cargo program is located at Johnson Space Center (Texas), while the commercial lunar lander, which is now a reusable rocket program, is run out of Marshall Space Flight Center (Alabama). Lastly, Gateway's NASA management is a bit all over – from Glenn to Johnson, to Goddard, and back to Kennedy Space Center. It turns out NASA's commercial space programs have not only grown in funding, scale and complexity. They've spread infectiously across most NASA centers too.

Looking for a pattern in all this won't find any just yet. Not that a data set of four (or five) should ever be expected to connect the dots. Commercial cargo program hardware tended toward the low hundreds of millions, whereas the commercial crew program *added a zero* to these figures, with hardware in the low billions. If you were placing bets, a lunar lander would tend in the lower billions too, as this is just like a crew spacecraft, or even simpler, as it does not need to come back to Earth. Except the low billions of a lunar lander, predictably of a scale of a commercial crew spacecraft, ended up being enough for NASA to contract with SpaceX for developing Starships. That includes a Starship booster with 33 engines. Better yet, this Starship will go to orbit like a Space Shuttle. And more tanker Starships (with their boosters) will refuel that first one, so it can leave for the Moon to get the lander job done for NASA. If the lunar landing bet were for an amount of NASA funding, you won! But if you predicted what that amount would be for, you would be *not even wrong*. (I speak from experience.)

There are more commercial programs in the wings for what comes after the International Space Station and for spacesuits. Yet, with the end of the ISS one day, the fate of NASA's two major operational commercial programs would be in the air. Moving commercial cargo and crew to within an entire commercial space station paradigm could be the start of unprecedented growth in private sector markets. But, on the other hand, it could also be scaling back with reduced NASA demand. Would commercial cargo and crew services to orbit be ready to leave the NASA nest and grow, or would there be a failure to launch? NASA gladly reminds us in programs like the ISS that we are all connected. So, we should not be surprised, with Russia invading Ukraine, if plans for what comes after the ISS do not survive first contact with worldly events.

Seeing that 16% of NASA is so different than what came before, an observer might be an optimist or a pessimist. The optimist can see NASA managing a mutual fund for space exploration. Congress directs NASA to spend on specific programs, with orders including who and where particular to a district. These old-style programs live alongside these new commercial partnerships where Congress provides funds with plenty of leeway. Here NASA - go do what you have to do. No one could promise who would build a spacecraft for crew or if NASA would choose two partners instead of just one. Like a box of chocolates, with partnerships, a congressional representative is never sure what they are going to get. This mutual fund has all the stocks, from low risk to high risk to some cash for liquidity on the side. Competition among programs, and a dose of healthy chaos, come with imagining and innovation. It helps much more than hinders, in the end.

A pessimist, though, might wonder if all this mixing-and-matching makes sense. Competition is good, an essential element of partnerships, but is all this competing among interests too much of a good thing? A mission to the Moon must have a singularity of effort. Having been tossed such an assortment of project types, is NASA supposed to MacGyver its way to the Moon? If we have a USB cable, a bottle of chlorine, and a table leg, maybe we escape from the jail cell? Or not. Because as much as the space exploration portfolio has lots in it, we should not automatically expect it's all we need.

No one could have predicted ten years ago just how diverse the NASA portfolio is today. Depending on your view, NASA's latest holdings are diverse, more robust than ever, or all over the place and uncoordinated when so much else is also in flux. No wonder "challenges" is a fan favorite of NASA-speak. We know that 16% of the portfolio is trying new things, public-private, commercial, and adaptable by design. So while we may not have the data to spell out how this all goes, we do have a good idea of who is ready when change happens.

SPACE BENEFITS, STEM CELLS, AND WHY WE'RE JUST GETTING STARTED

Early September saw some good news in the space sector, but not of the usual sort that quickly goes viral. The University of California San Diego received a gift of $150M to fund the Sanford Stem Cell Institute. Their valuable work with stem cells already includes years of research in Earth orbit. Yet news like this easily gets lost in the fray among the announcements of new companies planning launchers or satellites, the big NASA rocket leaking, or some other rocket exploding. Add to this the videos, any engine firing is a must, and the time to take in what's going on is over for the day. Move on, get back to work. Petri dishes have a hard time competing with fire and thunder.

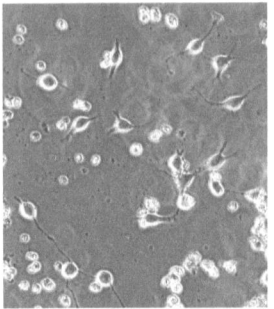

Microscopy image of Human Neural Stem Cells cultured for 45 days. Credit: UCLA/Dr. Espinosa-Jeffrey

For years after the Shuttle program ended over a decade ago, it was popular to reminisce about accomplishments. There was a lot of fire and thunder and 135 launches, with 306 men and 49 women who flew to space, and so on. Like a tour guide at Kennedy Space Center, the idea was to inform and to impress because who wouldn't marvel at how fast a Shuttle engine could drain a pool. One presenter gave a warning before a particular slide, in a time before anyone would have said "first, a trigger warning." "*Some people*" feel there is value, he said, in "*merely* having people in space." So here was the number of days NASA crews spent going around in low Earth orbit in Shuttles. 8,337 person-days in space, to be exact.

The point was made, and the accomplishments were impressive. But how such a jumble of numbers mixes activity with achievements is not lost on an audience. We need context.

As NASA deals with hydrogen leaks again, people will remember that time decades ago when the Shuttle too was plagued with hydrogen leaks. I pulled the night shift at the launch pad as technicians changed out a leaking 4-inch hydrogen line connecting the tank to the orbiter. The whole team felt a sense of urgency and a common mission, to fix what's broken and do it right, so we can launch again soon. This was one of those nights, not as common as they should have been, when it seemed everyone wore the same badge. It was two in the morning, and everyone was wide awake and determined to get the warp drive back up and running.

After beginning double-duties with advanced projects, eventually my full-time job, those years in the Shuttle program proved invaluable in revealing a picture that is still developing today. Transportation to space was important, of course, as a means to an end. No means, no end. Some advanced projects began with the end in mind, such as space based solar power, working backward to the means, routine, much more affordable transportation to and from space. The distinction was clear, why go to space, why live and work there? For the tremendous benefits we once had only an inkling of, but proving out with every visit.

Similarly, as NASA and our partners built the International Space Station, a facility in space, we could not forget this was infrastructure, a place of work. With no intention to take away from the immediate benefits of the means to travel to space, or the means to stay there, still, we could rightly tell anyone you haven't seen anything yet. *We're just getting started.*

Also, this week we saw the Space Council emphasize the benefits of being in space. Addressing climate change, NASA will develop an Earth Information Center to equip decision-makers with the information they need to respond. Local areas (like Kennedy Space Center) already have Geographic Information Systems. It's time we have one of these systems for the whole of Earth. There is none greater than keeping our home in working order, habitable, and safe when it comes to benefits. Pakistan one-third underwater is no longer a black-swan event, any more than record temperatures around the world, historical glacier melt, or wildfires across the west that are worse each season.

Right along, NASA Administrator Nelson talked about the benefits of stem cell research in space and NASA working with the makers of the cancer treatment Keytruda. There is an awful lot to learn about materials, proteins, and monoclonal antibodies, if we study them in space. There is a world of problems here on Earth held back by gravity but which, if resolved, promise, dare we say the C-word, *cures*. At least for the moment, the Petri dishes usually lost in the news were front and center. The Space Council members emphasized why we go to space and the value of performing research away from the effects of gravity.

Platforms and devices used for stem cell research under real and simulated microgravity conditions. A, The International Space Station enables long-term cell culture in real microgravity. B, Custom-made random positioning machine (RPM) for the cultivation of mammalian cells under simulated microgravity conditions.[20] C, The rotating wall vessel (RWV) is another type of microgravity simulator suitable for suspension cell cultures or adherent cells on microcarrier beads

From the Journal of Stem Cells Translational Medicine, "The effects of microgravity on differentiation and cell growth in stem cells and cancer stem cells," by Daniela Grimm, Markus Wehland, Thomas J. Corydon, Peter Richter, Binod Prasad, Johann Bauer, Marcel Egli, Sascha Kopp, Michael Lebert, and Marcus Krüger, 2020.

To get away from gravity (mostly), we have the International Space Station, a capability that has hit its stride. Yet NASA knows the ISS won't last forever, and more so, it's not a commercial production facility ready to crank out in volume. The ISS is a research site. So, NASA is helping commercial space stations in the works. It is not a stretch to think that a principal benefit of commercial space stations won't be serving a few astronauts here and there as a service, as research never ends, or lowering costs to NASA. Instead, the benefit will be in the products manufactured on these new stations that you or a family member might need one day. It may be an infusion to treat cancer or Alzheimer's or making crystals that show us the exact structure of the next pandemic virus and how to attack it.

Space capabilities and the benefits these will make possible do not stand still any more than solar panels, medications, or the knowledge of our home planet Earth. It happens home, health and humankind are the best of reasons to take risks, explore, and learn along the way.

Wrapping up the night, now near morning, at the launch pad years ago, the team having installed the new hydrogen line, there's a catwalk rarely seen along the back. Suspended forty feet above the deck, it was quiet enough to hear the waves coming in at the shore. The night sky was crystal clear, with more stars visible than usual, even with the brilliant lighting at the pad. We knew why we did what we did, as does anyone in the current space transportation business, and soon the commercial space station business too.

A CHECKLIST FOR COMMERCIAL SPACE AND NASA

A paper of mine was published in 2021 in the New Space Journal, "Ingredients and Anticipated Results for Characterizing and Assessing NASA and U.S. Department of Defense Partnerships and Commercial Programs."

Yes, that's a mouthful.

I often write about what's next for NASA, the commercial space sector, and how these must move ahead together. My favoritism for graphs and data shows, but these are a means to an end. My enjoyment is writing about the meaning and possibilities behind the numbers. This paper is about the latter, qualities, not quantities. The "New Space" journal is where I ask what best describes our "new space" projects?

In a nutshell, certain ingredients characterize NASA and U.S. Department of Defense partnerships and commercial space programs. The ingredients are obvious features, if not easy to pin down.

For partnerships during development, a "checklist":

1. Is the government committed to buying the product/service once available?

2. Does the government invest in many partners in very early development phases?
3. Does the government invest in moving at least two partners toward operational capability?
4. Did the private partner invest significant dollars in the project?
5. Does the government payout funds only at tangible milestones for results (not activity)?
6. Will the private sector retain ownership of the product/service?
7. Does the private sector partner demonstrate a commitment to a non-government market for the same product/service?
8. Is the government management office relatively small?
9. Has the government enlisted a standing, outside advisory group?

Once a project development finishes (if it does), the word "commercial" applies. (Though outside of contracts and formalities, the words "partnership" and "commercial" are used interchangeably.) So again, a checklist can tell us a lot.

(1) Will there be at least two providers (and more competition along the way)?
(2) Is the government buying at a "firm fixed price"?
(3) Is the partner attracting non-government business?

There are reasons all these ingredients get baked in the pie (the "anticipated results" for all this.) One is how traditional "cost-plus" contracting is more often than not "heads I win, tails you lose," from the private sector's point of view. But there is a lot of rhyme and reason here to these ingredients.

So rather than "contractors," we move to "partners," and NASA, the DOD, and the private sector sink or swim - together. An anticipated result is, yes, more results. We move away from paying to cut the grass by the hour, which goes slowly. Very. Slowly.

But what does this checklist mean, and what use is it? When you look at these ingredients vs. projects, you see patterns. We can look back to SpaceHab, X-33, EELV (now NSSL), Commercial Cargo and Crew to ISS, and XSP. NASA's Commercial Cargo program had a lot going on, from rockets to spacecraft.

The more these ingredients you can check off, the more likely your project will succeed. Not getting canceled, success. Meeting many of your initial goals, success. Wildly exceeding goals? *You probably checked off the whole list.*

Looking forward, what does this mean for commercial programs in progress? We have CLPS, a crewed lander, spacesuits, private stations, a Gateway, and launching cargo (again, now to deep space.) Read your ingredients list, and you'd be surprised at what you see. What does the checklist say about the prospects for these partnerships and commercial programs? At the least, the "checklist" tells you where to pay attention and ask questions. It also says where to be concerned or where things seem on track.

Not everything is about numbers, as I point out in the paper. I dive deep into numbers elsewhere and often. This is a different tack, like the saying, because not everything that counts can be counted.

NASA: MAKING MARKETS, NOT ROCKETS?

There is an old joke in NASA, "a million here, a million there, before you know it, you might have real money." It's probably a line in any business grown large enough to develop an unhealthy disrespect for money. Yet our more serious discussions enforced the same idea. Could NASA nudge industry this way or that, in the dry world of standards or the exciting world of launchers and spaceships? Probably not, argued most, a long time ago. In the big scheme of things, NASA was just too small. All we could spend was a billion here or a billion there, not real money. We were small fish, and the big fish were over in the Defense Department, or they were our prime aerospace contractors. At best, NASA could be a bee stinging a giant, an annoyance, and not much more.

If that was then, what about now?

Recently, the world saw how a giant bit by a bee could be more than just annoyed. The headlines for a while were about the short-lived government of Prime Minister Truss in the UK, axed when a financial crisis followed the announcement of a plan to cut taxes and borrow the difference. We could forgive the PM's number crunchers for being clueless to the chaos that would follow over a mere 45 billion pounds spread over five years. For a little perspective, that's under $11 billion yearly or only 38% of NASA's 2022 yearly budget. The UK finance folk must have missed the joke. The plan was real money, after all, not for the small amount itself but for what it affected.

The money with a "t," as in a trillion, reacted swiftly for the worse. There was talk of implosion and bankruptcy in the vastness of UK pension funds, bond markets, and banks. Things start to break when a giant bit by a bee jumps around in the China shop.

Part of NASA gets the joke now too, how little bits of money can have effects far out of proportion to the amount everyone debates. But rather than threaten to bring down a national economy, parts of NASA have applied relatively small amounts of funding to build new economies. This ability to create or destroy wealth comes from leverage and interconnectedness.

If small fiscal changes can bring down a government, evaporate billions in wealth, and shrink markets enough to notice, we can imagine the opposite. Take (relatively) puny amounts of cash, so millions bring about billions. Investments do this all the time. Now let your imagination run wild and ask what billions might do. Instead of saying, "NASA spent," say, "NASA invested," *and say it with meaning.*

In 1994, a US aerospace industry group looked at a version of this investment effect. If costs drop to put something in orbit, at what point do significantly more people show up who would never have launched anything? The idea was "elasticity," and the report returned at a daunting 661 pages (the full version is here and the short version is here.) Ask a complicated question, get a complicated answer. The simple version was the "hockey stick" non-linear jump in demand was at a cost way below where everyone was at the time. The matter was getting to these low costs, and how NASA's investments play a role.

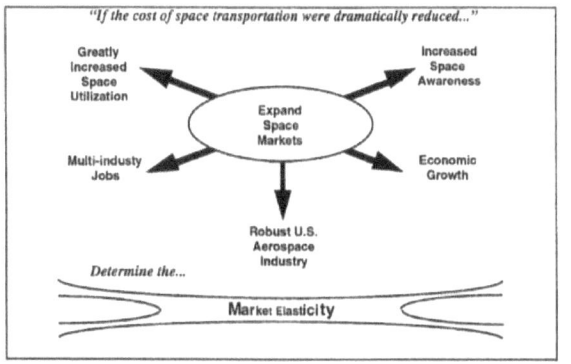

"Market Elasticity," from the 1994 Commercial Space Transportation Study.

NASA tested this idea in the real world many times. There were X-vehicles and focused R&D, and even US Defense Department launch was going "commercial," for a time. There were failures and mixed successes. But if at first you don't succeed, with the Shuttle soon to be retired, NASA decided it would try again. This time the goal was to create a commercial market for cargo to the International Space Station. NASA would make a market, not rockets or spaceships.

In 2006, SpaceX became one of two private sector partners investing alongside NASA to build new ships to get cargo to the ISS. Never did some hundreds of millions do so much good. Without NASA's investment, and later awards buying rides once available, SpaceX might not exist. This means all the commercial satellites SpaceX lured away from Russia or Europe's Arianespace might never have launched from the US. This means the multi-billion-dollar Starlink network might not exist. This means our talk about one day having launches every week would still be a goal, not a reality, as it is now. Ditto for reusable rockets. By hook or by crook, NASA became a market maker and not a bad one at that.

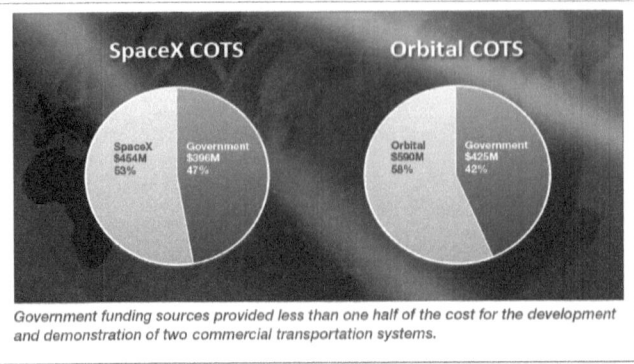

Government funding sources provided less than one half of the cost for the development and demonstration of two commercial transportation systems.

Do you have a story about seeing the future and buying Amazon stock at $10. NASA has an even better story. "Commercial Orbital Transportation Services, A New Era in Spaceflight," 2014, NASA.

If investments are about a new thing, market elasticity is about the shape of the business around it. Did new customers show up to take advantage of such low SpaceX prices? If Starlink counts as an idea no one would have done at previous costs, then yes. But only SpaceX gets SpaceX launches at cost, not price. By price, the data is not clear that a notable hockey-stick uptick effect is apparent from lower SpaceX launch pricing. Though in small launch, customers appear to have shown up that arguably might not have been around at all otherwise (and SpaceX is capturing those customers too.)

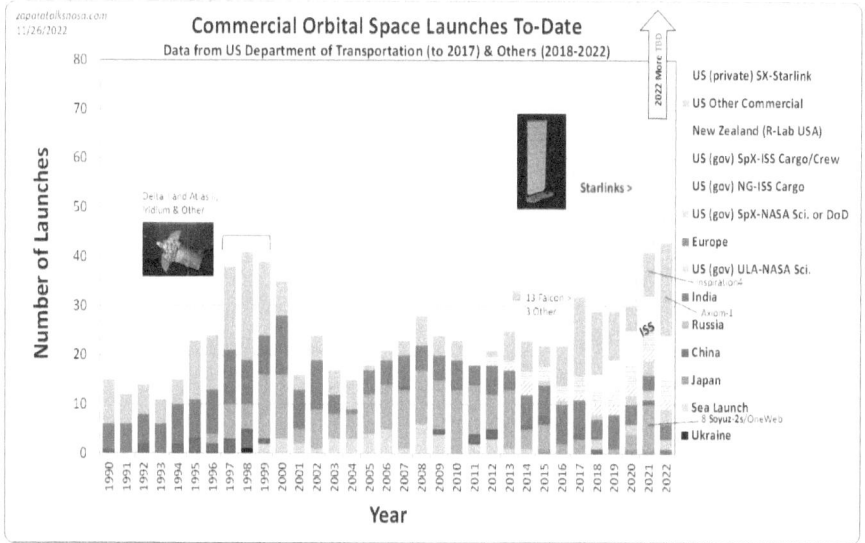

Subtract Starlink and there remains a slight uptick, but not quite as dramatic a "hockey stick" effect from the lower launch prices available from SpaceX.

There are other parts to these NASA moves where things have not fared as fantastically. Antares and Cygnus were also an outgrowth of NASA investments with industry toward developing a new way to get cargo to the ISS. Yet the price for these, though attractive to NASA relative to Door-Dash using Space Shuttles, has not been attractive to anyone else. There have been no non-NASA ISS Antares launches.

Similarly, NASA's investments in commercial crew providers have resulted in only two non-NASA flights (to date), both by SpaceX. Boeing's Starliner has no non-NASA customers lined up. Promisingly though, more non-NASA crew flights are planned (again, by SpaceX). For crew, it took billions above the hundreds of millions for cargo ships to develop new spacecraft. Add a zero to the monies, and get life support and amenities for people. And the zeroes keep growing to where we see Spacesuits as rentals, but still on the pricey side. The why and at what of all these will eventually make for some curious forensic accounting of NASA's investments, a subject I am familiar with (for cargo).

NASA partnerships are no longer about a million here or some millions there. We've seen how leverage can turn millions in NASA into billions elsewhere. As Archimedes said, "if you give me a lever and a place to stand, I can move the world." That is so for NASA too. But as we saw with cargo to the ISS, it's not the amount of money that matters. It's interconnectedness. Just what are we applying the lever to? It's not about money alone, and too much may even be counterproductive, forgetting the bar for pricing is way below current prices. (High volume will be essential.)

Admittedly, it's not easy connecting NASA's needs with the needs of others for markets that may not yet exist. But when NASA makes the connection, they have a lever that can move the world.

SPACE, PLAYING THE LONG GAME

If you follow the space sector, and maybe even if you don't, the unavoidable impression is there's so much happening fast. Space stuff and that AI shows up at every party. The days when only an occasional Shuttle mission, Hubble picture, or a Mars rover made headlines are in our rearview mirror. Today, it's always something, from Starship to a stuck antenna or a failed lunar lander. It appears someone births a new space project promising to change life on Earth as we know it, every week. So much drama, and we love it.

There's also catching up with old friends. We hear the new US Vulcan and the European Ariane 6 rocket debuts are delayed. These rockets, like too many others, are close to joining the Blue Origin *are we there yet* club.

There's sheer ambition and innovation too. To make you feel like getting to orbit is easy, try returning a stage from orbit. For this we have Stoke Space. And even as Virgin donates its organs in bankruptcy, Astra and Firefly's nascent small launch offerings still have a pulse.

For some non-rocket rockets, going with brute force, we have two companies when one alone would be surprising. So SpinLaunch and LongShot Space are both building facilities to hurl cargo to space, one using a centrifuge, the other gas guns.

For a welcome dose of hypersonics we have Hermeus building a Mach 5 airplane. Is this part of a long and winding road to spaceplanes? After Mach 3, it's no moving parts (almost) – the reason future historians will wonder why we took so long to figure this out. Be kind. We are trying.

Uncertainty has never been so exciting, though some of these projects (okay, let's say *most*) will not make it. No matter, all over the world, not just in the US, getting to space or having something in space sending back images, or providing broadband, is turning old, tried, and true business models on their heads. Now it's Europe proposing a Starlink competitor – IRIS[2]. Tomorrow it will be something equally significant. It was time.

Yet in all this, we forget to ask – *why*?

Suppose you are on the rocket or spacecraft side of that tyrannical equation about mass and gravity, as I was. In that case, confessing to a certain obliviousness outside of transportation is natural. NASA's spaceflight budget reinforces this fixation, as most dollars go toward getting to space or simply staying there. A big rocket? Transport to orbit. A capsule? More transportation once you are in space. A future lunar lander? Transportation again, to get as far as the surface of the Moon. Plans for private sector space stations after the ISS? Staying in space, a facility. (For some context, the ISS R&D budget this year is only $306M, or 9.4% of the ISS budget. The rest goes toward getting to it and keeping it running – transportation and infrastructure, not use.)

The private sector worldwide continues with this lopsided emphasis on transport. We'll all worry about what we do at the destination and what we carry back and forth after we figure out the cars and the trucks. Let's first figure out how to get to space and return easily, cheaply, and often.

In a presentation some years ago, but not the first time, a presenter looking back at the Shuttle program asked the audience to think of people in space as an end in itself. Forget, for a moment, the experiments. Put aside the satellites we deployed, retrieved, or repaired. Fixing the Hubble Space Telescope? Not easy to graph, though the pictures were a blast. Also, put all we learned over on another pile for later consideration. Now focus on the people who went to space. NASA went from a narrow band of US astronauts in a previous generation to hundreds of men and women of diverse personal, national, and professional backgrounds going to space.

That time three astronauts on a Space Shuttle grabbed a misbehaving satellite - with their hands. STS-49. Image: NASA.

In this view, the road trip was the point. With the Shuttle era, the arrow on the graph pointed in the right direction – more people going to space more often than ever. There was the science, of course, and the learning to live and work in space, paving the way to build an International Space Station, now with people aboard it continuously since 2000. But here, the means held value regardless of the result.

Down on the ground, we knew getting to space was part of a progression. Routine transportation to space is a means to ever-growing numbers of people continuously in space. And more facilities and people in space were a means to deliver benefits for everyone back on Earth.

In 2023 Keytruda became the highest-revenue drug worldwide, a monoclonal antibody effective against an increasing list of cancers. As the wording suggest, this antibody does what antibodies from your immune system do – they go after a target, here cancers. ISS research has shown that "conditions producing crystalline suspensions of homogeneous monomodal particle size distribution (39 μm) in *high yield* were identified." In other words, a costly cancer drug could be turned into a cheaper, longer-shelf-life subcutaneous injection (it's currently delivered by infusion, an "IV.") This is just one example of the "demand" that would need a "supply" of routine, low-priced transport to space, and low-priced labs and manufacturing facilities too.

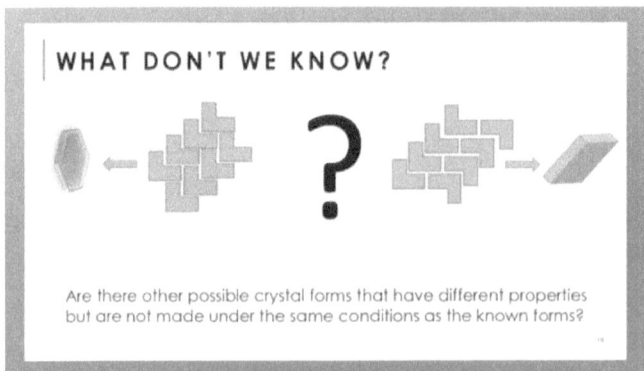

"In Space Production Applications with a Focus on Crystals," Dr. Savin, 2021, NASA Commercial Space Lecture Series.

The path from getting to space to staying to direct, everyday products and services that benefit people worldwide has not been easy. Satellite communications, imaging and sensing, and many indirect benefits have been with us for decades. Imagine the results if getting to space and staying there were much more accessible. Beyond broadband anywhere and anyone ordering satellite images lies the unimaginable. New materials? Medical advances? Cures? AIs helping point us in the right direction for all these? (Imagine better protein crystals made in space, antibodies, with AI more quickly discerning structure from the improved diffraction grade crystals.)

As NASA looks beyond the ISS, there is a huge opportunity and grave risk. On the one hand, the ISS could end as we see many private space stations in orbit and more coming. Product is shipping back and forth from Earth and public-private orbital factories. NASA declares victory, its investments in Commercial LEO Destinations and In Space Production and Applications transforming the space sector - again. Here, for a twofer, the simple view of more people in space, more often, also screams success. It's no leap to say ever more tangible results likely follow.

But imagine an alternate scenario, not rosy but quite possible too. Here the ISS ends around 2030, and at best, a small private module in space gets business from the occasional NASA astronaut. Occasionally some tourists drop in too. By the numbers, even after adding in NASA's lunar mission here and there, fewer people, less often, are leaving Earth than during our Shuttle and ISS eras. The arrow on the graph is pointing the wrong way.

Space is the ultimate long game. After going to space often, at great cost in human life, with the ISS we set up shop and stayed. Transportation is clearly a means to an end, the getting to space and back. Staying in space was another monumental step. At the risk of repeating myself, which I'm told will happen if you blog long enough, the exploration of our little solar system (and beyond) must not lose sight of the plot. The R&D, the results, the product that one day says "Made in Alpha" could become the MacGuffin of space projects. Or try, try again, and try what has not been tried, as all the latest efforts are doing, *and* keep our eyes on the prize – the wonderful everyday benefits ahead for people here on Earth.

EUROPE, ESA, THE EU, AND THE SPACE SECTOR - WHERE TO NEXT?

It's the mid-1990s, attendees lingering in the expansive lobby as the day's conference presentations concluded, everyone glad to make new acquaintances and hand out business cards (remember those?) Among the introductions and salutations, there is a representative from Arianespace as we wander over to a corner and some overstuffed chairs to "talk shop." We focused on work much too quickly after the pleasantries about Paris, flights, and a restaurant recommendation. Of course, proceeding cautiously on any topic of food, as any local suggestions could be akin to proposing Twinkies to a dessert chef. As for the wine, that's definitely from California. It took only fifteen minutes to nearly cause an international incident. But it was about the Shuttle and Europe's successful move toward commercial space – not the wine.

My presentation was roughly version 0.5 of a presentation I would hone and improve over the coming *decades*. We were slowly unearthing what the Space Shuttle cost in all its gory detail, though I had only started to understand the depth and never-ending nuances in each number. There were the numbers that lied, the ones with an important story, and the ones generating more questions than answers. Much more meaning would come, just not all at once. This is near the start of my story.

At the time, I may have appeared to be an up-and-coming critic of the Shuttle, where "critic" loses the sense of looking at things analytically, objectively, and "critically" to improve. The term critic too often conveys only opposition. I saw my critique as part of learning and finding solutions to do better next time, not the latter about merely pointing out problems. Many fulfilling years awaited me as part of a team preparing and launching Space Shuttles.

My Arianespace acquaintance was familiar with my topic and the Shuttle's shortcomings regarding affordability. This may have been the first time I heard the phrase "commercial" operations – "Arianespace is an efficient organization, being commercial, while the Shuttle is another bureaucratic, inefficient, government-run enterprise." More or less, as memories go.

At the time, Europe was leading the pack in our space sector. Whether they were the Space Shuttle or the Air Forces (today's Space Force), US launchers were not competitive. By law, the Shuttle no longer flew commercial payloads since after the loss of Challenger. The Air Force's stable of rockets was dated. The joke about Titan rockets being the Air Force only launched these once they rusted solid to the pad, or the weather turned to gray soup with no visibility, or both. The Delta II expendable launchers were workhorses, but the US Defense Department wanted bigger. And the Atlas rocket of the time came scared with a legacy fixation around reliability and mass to orbit, not competitiveness.

Europe, though, had gone commercial, with a highly competitive posture putting the first nail in the coffin of US commercial launch. The trend was there already. Soon, the nascent low Earth orbit constellations, like Iridium, which threw business the way of those US Delta's and Atlas's, would end in a dot.com bust. And a final nail, as Russia would be allowed to launch private-sector western satellites as "midnight basketball for Russian rocket scientists." Let them launch some satellites for cold, hard, western currency – this should keep them from getting into trouble selling their know-how elsewhere.

So much has changed since. But here I was at that conference, and parts of this future are yet to be written. Still, Europe already had bragging rights at this point in the story.

In response comes another development, the US recognition by the Defense Department that its rockets must, like Europe, also "go commercial." The Evolved Expendable Launch Vehicle program emerged from this effort, leading to today's rockets – the Atlas V and Delta IV. These were bigger and better but more expensive and less competitive. Ironically, the Defense move toward commercial rockets and competition led to a monopoly formed by both companies (after a scandal with Boeing stealing Lockheed documents.) It would not be off the mark to describe this as the fox chewing off three legs only to find it's still in a trap. The monopoly did not end until a decade later, with some cajoling by SpaceX.

Yet this outcome, too, is years away that day in the hotel lobby. As someone intimately aware of the Space Shuttle's flaws regarding affordability, or the lack thereof, it could have fallen to me to say "inefficient" – as only the start of a critique. Scouring for all manner of data, it also came from the Air Force about their rockets (and aircraft, too.) In the words of some of my new Air Force counterparts, "Compared to what's going on over here, NASA is downright efficient by comparison."

In short, Arianespace was pulling in business, while the US launch sector was not. The future of space launch appeared commercial *and European.*

Europe had created a lean, mean corporation with every incentive to be affordable. This was not only to serve the scientific payloads from the European Space Agency ESA but also for the additional private sector payloads they could attract. Everyone wins.

This all sounds familiar years later.

The US commercial launch breakthrough fell to NASA for reasons that remain obscured by the fog of war. Just what did NASA do? The NASA "COTS" program would pay to deliver cargo to the International Space Station as a service. The usual narratives describe parts of this elephant, depending on who grabbed what.

If you see the world through contracts, here is the moment NASA said enough is enough. These are rockets. This is a simple spacecraft, Apollo style. This is non-human cargo delivered to a point in space through an airlock. It's supplies, lunch, water, and well-packed gadgets. There is no reason we can't just sign on the dotted line for a *firm fixed price,* as all there is to do here was long ago known.

But the elephant is so much larger than its trunk.

NASA would pay out on those contracts for more tangible outcomes than usual. At milestones showing progress, a check would be cleared, a partial check. This shifted expectations away from activity and toward results. It was no longer enough to try and cut the grass or ask to be reimbursed for the hours spent cutting it. Or the hours planning to cut the grass or trying. When you are done with the north forty, give us a call for that partial check.

The partnership between the public and the private lies in the public paying and the private performing. Better yet, someone must have been on to the previous decades of flawed contracting with the best intentions. Here, selecting the "low bid" devolved into choosing the "best liar." Now, to keep everyone straight, a couple more nuances were added even after a partner wins a contract. There would be two partners, at least. Better yet, NASA showed early on it would not be shy about ending its relationship with one partner and seeking another – the erstwhile Kistler switched out for Orbital (later Northrop.)

And more – while the government would commit to

buying services once operational, it would not commit early to anyone, not even the first partners. Keeping options open made it clear – did we say results?

For all this to succeed, other necessary ingredients followed. The government management office would be smaller. Not a smidgen fewer government personnel than usual overseeing the cooks in the kitchen. Not like ten percent less. Small as in *tiny*. Tiny by comparison to the usual way of doing business. In the Shuttle program, fifty or a hundred people would have been called "a small team" for a study, for a minor change to the Shuttle (that likely never got implemented anyway.) Eventually, when sending a crew to the ISS using the same commercial approach as for cargo, this would be the scale of the entire NASA team. As we learned in the Shuttle program, in the worst way possible, the more signatures appear on a piece of paper, the less it's worth.

All this is about setting the stage until the arrival of SpaceX – leading the pack among partners, including Orbital (now Northrop) and Sierra. After all is readied, the show could close on the first night if the performers are dismal. Fortunately, this was not the case. Long before NASA took a risk with a new way of doing business, the space sector business environment was also changing. Know-how had spread, attitudes shifted, and new notions for developing and operating launchers and spacecraft now seemed credible, so worth the risk. Private sector investment saw an opportunity (not only in launch but in entirely new approaches to satellites too – like Planet.) For reasons Europe might also ponder on the success of Arianespace in the 1990s, the performers not only showed up, they wowed the audience.

Table 1. A checklist for NASA partnerships and commercial acquisitions

Up-front Ingredients	
1. Commitment – the government's intent to buy when operational is clear	☐
2. Competition – the government invests in multiple partners in the very early project phases	☐
3. Competition – the government invests in at least two partners in the development phase	☐
4. The private partner invests significant $ in the project	☐
5. The government provides funds at progress milestones/results (favors "what" over "how," 1 of 2)	☐
6. The private partner owns the system	☐
7. The private partner demonstrates a commitment to a non-government business plan	☐
8. The number of government management/personnel is small	☐
9. The government enlists a standing/independent outside advisory group	☐
Operational Ingredients	
10. Competition – the government procures from at least two partners when operational	☐
11. The government procures the capability/product at a "firm fixed price" (favors "what" over "how," 2 of 2)	☐
12. The private partner attracts non-government business at least on a par with its government business	☐

"Ingredients and Anticipated Results for Characterizing and Assessing NASA and U.S. Department of Defense Partnerships and Commercial Programs" New Space Journal, Zapata, 2023.

But back in the lobby after that conference which now feels like a retelling of a curious story during the Ming dynasty, my brain would not tolerate someone else labeling the Space Shuttle as one more inefficient government-run enterprise.

I could do this. Others in NASA with a shared goal of improvement were also free to speak frankly.

But at that moment, I switched gears to defend the Space Shuttle program – my program - vigorously. We could retrieve satellites and repair them. With teamwork and courage, our resourceful astronauts could grab satellites - *with their hands.* Jupiter, Venus, and probes and all we accomplished or were going to soon. I may have gone slightly overboard – to judge by the awkward silence that followed. Soon after, I felt I would get a call about the complaint lodged at the French Embassy. In retrospect, this conversation was tame compared to those to come.

Pressing fast forward, today, the new model Ariane VI pricing is on target to compete well - against yesterday's rockets, not SpaceX.

Report after report encourages Europe to do much as the Americans have – creating an environment for private sector investment in space, simplifying their "complex governance system" (we call that paperwork and meetings), and overall finding and risking capital, public and private. If the formula for success were known, probably there would not be yet another report. The challenge is also out there to offer a path forward for European space - that is legible rather than hundreds of pages of re-stating problems. Making issues less clear by page a hundred and twelve is always a fantastic accomplishment.

It's easy to sit back and say what goes up must come down - in competitive industries or Earth's orbit. Adversarial perspectives come straight along, and it's win or lose and who's on top. That was me, too, long ago in that lobby. More Zen-like, with time, we might resign ourselves to the swing of the pendulum, compelled but at peace tracking its course.

Or is there more?

Europe and others in the space sector will find their unique paths. There will be no lack of committees, reports, and recommendations. Does this end with an Airbus for every Boeing, or silos of capability, east and west? This seems quaint, linear, and limiting.

In a different ending, everyone is learning from risks, experience, and each other. We should look forward to everyone emulating but also adapting and improving on what others did before. It's what this vast enterprise to expand human presence beyond Earth just might need to go over the top. Beyond competition, it's everyone in a learning loop. The attitude over wine at the gathering is *the more the merrier. Rather than* looking to leave others behind, it's everyone enjoying moving forward together.

A PICTURE WORTH A THOUSAND WORDS - FLIGHT RATE, NASA AND SPACE EXPLORATION

We needed launches. Lots of launches. That much was clear, even if how to get there was not. It seemed it was always the same meeting, about a launcher, real or imagined, a Shuttle upgrade or some vehicle post-Shuttle. Perhaps the rocket was expendable, the big dumb booster, or maybe it was reusable. Perhaps it was not entirely a rocket but a spaceplane where the engines and the airframe were better thought of as a whole. Predictably, the reliability analysis was at the end of the day, following from what came earlier. My field of operations was usually nearby.

This left the impression that design decisions were first and consideration of consequences came later if time permitted. Like reliability analysis, operations could be mistaken for ill-defined and tentative results stemming from technology seen as real and final. Yet inside operations were launch processes, and depending on how often a launch was likely, there were launch rates, the purpose to it all. Similarly, inside those reliability numbers was technology maturity, a one-off now that might one day be common. And all this was connected.

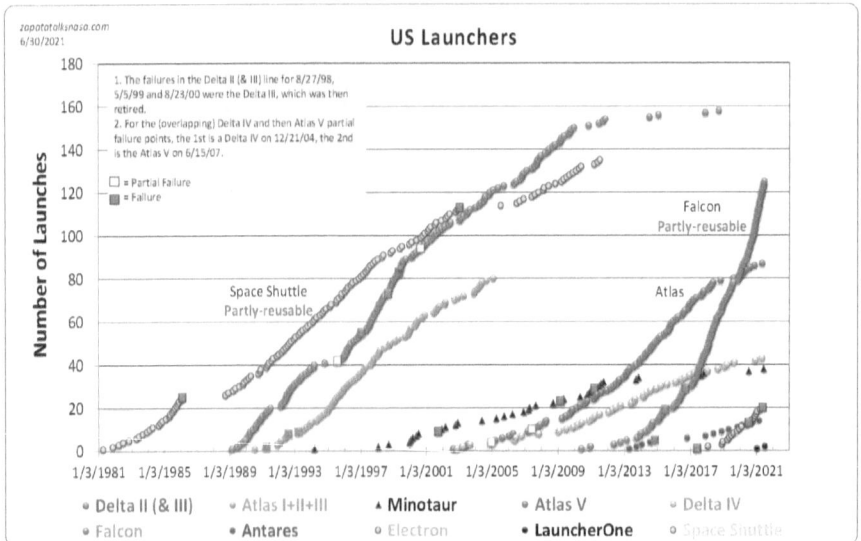

US launches over time and by launcher, from the first Space Shuttle flight in 1981 through the LauncherOne launch June 30, 2021.

This story begins with how complex projects like to break themselves up into a million tiny pieces. The notion is that the small pieces are much more manageable because they are limited. Project management must have birthed the military strategy of "divide and conquer." Still, well before it all meets up again in the real world, on the plant floor, or at the launch site, an analyst with a penchant for putting the pieces back together does so, at least on paper. We can't help but take a peek right now to see how it might add up in the future or how it might not. Having checked trees all day, someone dives in to talk about the health of the forest. The picture may not be what's expected from the earlier walk in the woods.

Plotting US launches by date and by launcher is not new. The data below, going back to the 1960s, including the Apollo program and the Moon landings, shows some impressive launch rates (the Atlas D in 1960). Those were different times, especially regarding NASA budgets, urgency, and the sense of national purpose. This inspired me to create the graph above with US launches since the Shuttle's first flight in 1981. There have been far fewer failures in the recent decades of spaceflight than in the first decade. Learning can do that.

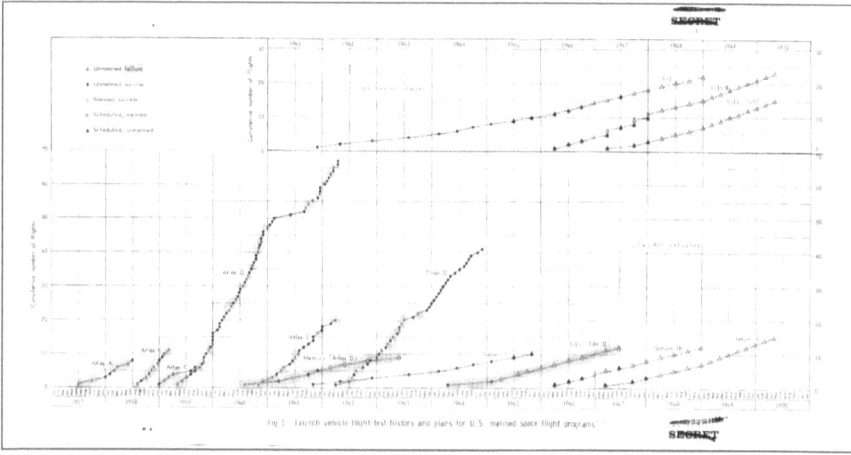

USAF ICBM and NASA Launch Vehicle Flight Test Successes and Failures. From "Apollo launch-vehicle man-rating: Some considerations and an alternative contingency plan," report by the RAND Corporation, for NASA.

It's simple enough to say, "Practice makes perfect." If only it were always a simple matter of how to practice. In launch-speak, this translates to "a high flight rate cures all ills," leaving the problem of how to get the high flight rate. Spaceflight debates also simplify this to a "chicken and egg" problem. A launch system, meaning people, must launch frequently to improve by learning, but can't because it is so expensive here and now. Reliability for rockets can never be aircraft-like until you've flown about as often. A variation on this problem is specific to reusable launch. Here a reusable system flying often can make up for a higher up-front cost, but the customers justifying that high flight rate do not exist because no launcher like that is available. ("No, you go first".) The expendable rocket version of the trap shifts this vicious cycle over into manufacturing rather than turnaround. Low-rate production of complex technology is expensive, dampening demand and making the low-rate production a self-fulfilling prophecy. However, for an expendable rocket, there is another matter - throwing away your airliner each flight will never be cheap, no matter how much you learn about manufacturing. If we've learned anything, it's to be reusable.

There is another hitch, too – carrying crew. Reliability and safety are two sides of the same coin, as parts that constantly act up on the ground can hardly be expected to behave in flight. Unfortunately, one way around these traps is unwarranted optimism. Take lots of flights, a billion dollars here and a billion dollars there, and before you know it, you assume the rest – that you must have made a massive dent in learning and making it all safer. The opposite is unthinkable. We want to believe massive expenditures reflect massive improvements, but sometimes, they merely reflect the difficulty of modifying a system carrying crew every flight. This led to the published Space Shuttle safety numbers just months before the loss of Columbia. The safety numbers below, from a Shuttle booklet handed out to employees in May 2002, proved to be woefully incorrect when reassessed.

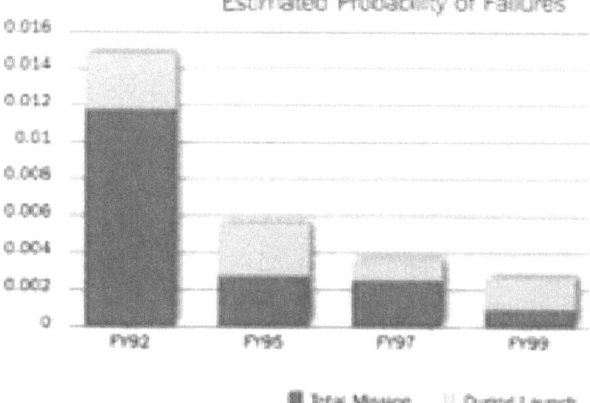

May 2002, 8 months before the loss of Columbia, from the booklet "The Space Shuttle's Second Decade, America's Best Gets Better" handed out to NASA employees and at trade shows. Publication NP-2002-05-019-JSC, NASA.

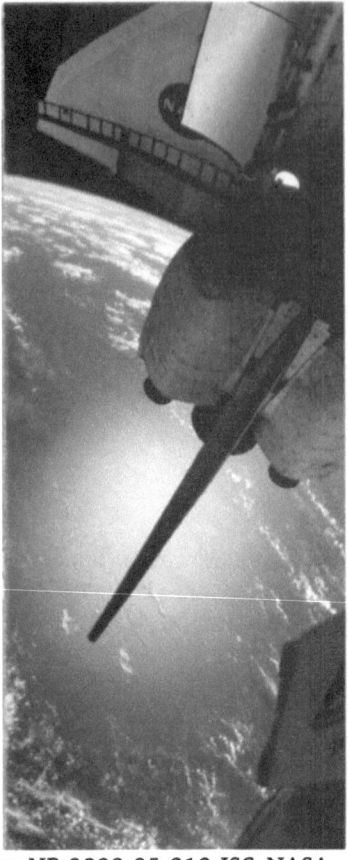

Publication NP-2002-05-019-JSC, NASA.

> "If you can't explain it simply, you don't understand it well enough."
>
> -ALBERT EINSTEIN

Simply put, a hundred launches, or a hundred of anything, is just a drop in a statistical bucket. It's never enough to chart a path to ever higher rates of launch that are increasingly cheaper *and* reliable. Or is it?

There are facts on the ground all around us that say these traps are not real. We are surrounded by complex technology people depend on with their lives, all quite reliable and inexpensive enough to be common. Planes, trains and automobiles figured out how to cross the valley from being scarce, expensive, and even dangerous, to being common, affordable and safe. A common retort here is the physics of getting to orbit makes these comparisons invalid (repeat mantra here, "space is hard"). And yet, we have the beginning of a steep climb in launch rates with the SpaceX Falcon 9. The dots are even atop each other, with no space in between. This is a good thing. Are SpaceX and NASA going around the traps we thought were always there, dooming getting to orbit to only glacially slow improvement?

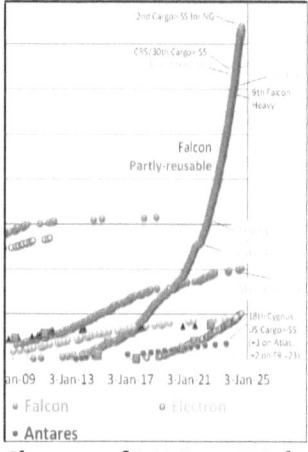

Close-up of 370 SpaceX Falcon launches vs. everyone else, as of August 4, 2024.

First, the NASA investment that resulted in the Falcon 9 and the Dragon cargo spacecraft could assume more risk, at first carrying cargo, not crew. Even better, backup options existed for delivering that cargo to the International Space Station. More risk opens the door to innovation. With innovation come new ways of doing business to be more affordable, attracting more customers, so more opportunities to learn. When the inevitable failure does occur (100 flights ago for Falcon), the improved tempo assures you quickly have "lessons implemented," not just "lessons learned."

Lastly, not enough demand? Provide it yourself with another innovation like a Starlink low Earth orbit satellite constellation of thousands of satellites. Demand gets the driver's seat finally, as the egg decides a chicken is a great way to make more eggs.

Times have changed. None of the growth and launch rates we have seen in Falcon would be possible based only on NASA demand, with every launch and payload coming from a limited budget. Ironically, none of the growth and launch rates we see on a system like Falcon would be possible without NASA, from up-front investments to buying services and providing technical assistance.

All this points to increasing space exploration that is ever more reliable and safer beyond launch in low Earth orbit systems and systems on the surface of the Moon. Here's to seeing lightning strike twice and again, adding more lines on the launch rate graphs, climbing even faster. Some craters on the Moon will help, too.

SPACE BASED SOLAR POWER AND NOT LOSING SIGHT OF THE PLOT

"The General doesn't like it," he said, because "he doesn't want to own the big, easy target that's the first thing destroyed in the next war." So much for what we might do together on Space Based Solar Power. This would be a short call. The idea of a massive power station floating in space attracted this kind of humor in some of our US Department of Defense brethren. The reply was reused for storing and transferring propellant in low Earth orbit. That station refilling rockets in space – also the first thing blown to bits. And here we thought the DOD didn't like reusability. If you find that funny, you're in aerospace.

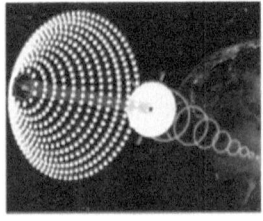

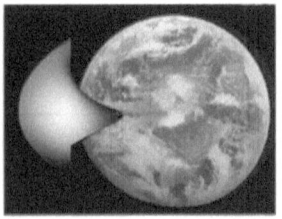

SPS Alpha
John Mankins, USA

Multi-Rotating Joints SPS
Xinbin Hou, CAST, China

CASSIOPeiA
Ian Cash, IECL, UK

Figure 2-1 Three leading Solar Power Satellite Concepts capable of providing baseload power
Concepts for space based solar power from the ESA commissioned reports. Credit: Mankins, Hou, Cash et al, as indicated.

Yet recently, the European Union *seemingly* came out in favor of space solar power. A "like" from the EU does not come easy. Here is hoping this gains momentum before they consult with a General. "Seemingly" should have an asterisk, as NASA also liked the concept long ago in the 1990s, spending real dollars and rounding up the usual suspects to figure it out, myself included. With time though, the initial enthusiasm of some did not translate into the action of many. By 2000, the idea mostly ground to a halt inside NASA. I put together my last formal report on this that year. The enthusiasm of some in Japan was also high then. Anyone at the time, seeing the reports, committees, and the press, believing this would take off, would have lost that bet. Still, the work of many people, with some hardware, if far outnumbered by analysis, all confirmed this was a promising idea for clean, abundant power. Ready for prime time is another matter – but time has passed, and why space based solar power stations are a good idea is much better understood now. The trick is not to lose the plot.

The charts flew by the screen in the small, dark amphitheater. The project compartmentalized everyone and their work, which is to say, we all performed small parts in a big play. Solar cells, efficiencies, check! Next presentation. Structures, mass, check! Finally, breaktime, before the charts flashing by give someone a seizure, as if we could take it all in like Data the android in Star Trek. This was one of many gatherings where everyone reflected on if their part fit with everyone else's parts. My role with a team at Kennedy Space Center was to see if very advanced space transportation could credibly fly often enough to build solar power stations in space. After much to-do, the answer was yes, multiple daily launches are quite credible if we work the right space transportation technology. We needed reusable spaceships, of course, designed to fly often. Our tack cleared up just what "designed to fly often" really meant. That was the good news from Kennedy, except even as all the parts came together, we realized something was seriously amiss.

※ ※ ※

As you drive between Orlando and Kennedy Space Center, as I once did five times a week, you will notice a power plant off in the distance. Two colossal towers make it hard to miss. Contrary to what a visitor may think, as some asked me over the years, this is not a nuclear plant. The laws of thermodynamics being what they are, the cooling towers are for a plain, old-fashioned coal-burning plant. This particular plant, the Stanton Energy Center, has the distinction of being the worse polluting coal plant in the US. The two coal burner units at Stanton each produce 440 megawatts of electricity. Not by coincidence, this is about what our space solar power station would produce back in 1999. A space solar power station must compete with terrestrial power sources, so it should generate similar amounts of electricity. It would have been easy to believe the green plant just replaces the dirty one, and off we go. Unfortunately, this is not the case.

Our planet is hungry for power. To measure global power, we use terawatts. These are units in the range of what the starship Enterprise has available, except they are a few hundred years in our future and remain science fiction. For now, we generate numbers like this between all of us on Earth, about 15 terawatts. We have the technological know-how to grow this power every year by the hundreds of gigawatts, adding more plants like that one you might pass on the highway, multiples of these - every year. We saw the basics of this race back in the late 90s. If we put up a solar power plant in space, someone else would add 40 times as much dirty power. That's the picture today when 18 gigawatts of coal power were added to the starship Earth last year.

If you have been in NASA long enough and want to do some good, you might try making a small dent in a big problem. For space solar power stations, the dent is about reducing the carbon dioxide we put into the atmosphere yearly. Preferably, make a big dent. Though on a personal level, if you're working on such a project, you might be forgiven for being greedy. In the long run, such a massive undertaking should, with other efforts, help us turn the curve worldwide, arriving at Year Zero – when we see global carbon dioxide emissions drop from the previous year. This would start a clear downward trend, not a fluke due to a pandemic. As our look at space solar power ended in the late 1990s, even making a small dent appeared unlikely, and not for lack of trying. Instead, we were marching uphill, as globally, a lot more power capacity would be built every year, and not of the space solar power kind. Our old and moldy charts, the ones we debated that day in the dark amphitheater, are still out there to show this. They remain pretty near the mark too.

AN ENGINEER'S JOURNEY IN NASA

STABILIZING CO2 LEVELS IN THE ATMOSPHERE

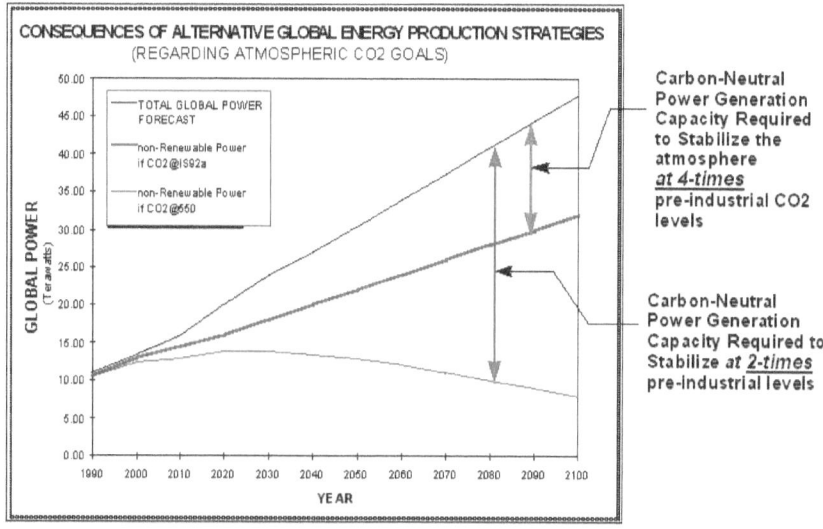

NASA Space based solar power circa 1999, view of global power growth versus stabilizing CO2 in our atmosphere via carbon neutral power generation. NASA.

* * *

Every year, dozens of coal plants go up around the world, and dozens more using fossil fuels like natural gas. China accounts for most of the new plants, while US coal plant capacity is in decline. These are the same kind of plant, if slightly more modern, as the US and Europe once had aplenty, ironically creating the wealth that makes it possible for us to move to cleaner energy sources today. In this global backdrop, building a relatively small solar power station this year, and another the year after, and so on, is hardly capable of making a difference. Part of this was so twenty years ago too.

Fast forward to 2022, when technology has a way of changing things exactly as we hoped. Back in the day (a phrase I use a lot since retiring), a solar power station in space came in at a massive 5,000 metric tons producing that measly 400 megawatts down on Earth. But a lot has happened since 2000. You can't keep a good idea down, or its advocates like John Mankins or Paul Jaffe. The EU reports bring us up to date with all the latest technology. Now we can plan for a station getting a respectable 1,440 megawatts to the grid and weighing in at only about 2,000 metric tons. That's much less weight, much more power, and all that means less cost.

In other words, technology for space solar power has not been still since the late 1990s, even lacking direct investment. You're likely to have a neighbor today with solar panels on their roof, everything electronic has advanced in leaps and bounds, and courtesy of SpaceX and NASA, we have Starships cranking out at a Starbase in Texas. As more good news, though the fossil fuel power added worldwide in 2020 was a sizable 60 gigawatts, it was slightly *less* than in 2019. More importantly, in the same year our world added a massive 260 gigawatts of renewable electricity.

Elon Musk is fond of saying space solar power won't add up. He attributes this to all the transmission inefficiencies all along the path from sunlight to electricity down on Earth. This is so he says, even if the solar plant in space sends down power all day. But this misses the point, even as it reminds us of the point. A solar power station collects energy all day, providing continuous power. So, while solar panels on Earth, and batteries at night, could end up doing much the same the world over, a mix of energy sources will be necessary across very different countries and situations. In space, it turns out, the sun never sets for a solar power station. This will prove priceless for rainy days or weeks or stretches when the wind that seemed so constant isn't.

As you drive to Kennedy now, you still pass the coal plant. It's slated to be shut down in 2027 and replaced with natural gas units. Up ahead, the sights have changed. As you approach Kennedy and near the gate, for the longest time you would have seen that staple of Florida tourism, orange groves. Coming up from the south, the required pit-stop meant dropping into the Crisafulli Groves store for a bag of perfect, delicious, if pricey, oranges. Those days passed, and the stores shuttered in 2017. The orange industry in Florida and worldwide has been under stress for decades. "Citrus greening" is only one of a host of threats orange groves continue to face. This creates "ghost groves," acres of orange trees that no longer produce many oranges. The root cause of most of these stresses is climate change.

Having once had a few citrus trees, it's sad to see them slowly wither away. Trees that once gave more fruit than you could give away at the office, or turn into smooth OJ that puts any store-bought kind to shame, produce less one year, then less, and finally hardly at all. A lesson learned, enjoy the oranges while they are still there.

※ ※ ※

By 2020, my drive to Kennedy had come to a stop. In this different world, telework already meant an infrequent commute. But then came the pandemic. After everyone went home to work, but before the vaccines, I found myself in route again to Kennedy. Zoom just wouldn't do. I went to one of our labs with special permission to meet with our research team. Near the gate, the site of an abandoned orange grove, workers were busy forming frames and rails and unpacking solar panels. Months later, when I retired, I saw the installation was complete, 74.5 megawatts of clean energy on 500 acres. It's all connected now like we only began to realize long ago.

The threads between spaceships, and solar power stations in space, the climate we have changed already and what change is yet to come, the land and the trees, and what we will do are all dots to be connected. It's likely *not* a General's job to connect these dots. Maybe it's not a problem small enough for NASA, a company and its investors, or the European Union and ESA either. Albeit, earlier this year NASA expressed renewed interest in space based solar power. Though a problem created by disconnected actions, a sort of tragedy of the commons, isn't always solved by even more disconnected actions.

But one night sooner than we think is possible, we might see some receiver panels laid out on land far away, softly lit up all around, with a notable bright spot in the sky that never moves. The power will be flowing to thousands of homes. The acres occupied will be less, with as much or more power than the terrestrial solar farm asleep nearby. To circle the installation, we might plant some trees, or an orange grove.

Florida Power and Light's new Discovery Solar Energy Center is a 74.5-megawatt solar site, spanning 491 acres at NASA's Kennedy Space Center in Florida. Image: NASA.

* * *

Part 6

PART 6 AI, NASA AND WHAT THINGS WILL COME

One of the first uses people found for AI when it bursts on the scene is to summarize. Give an AI a long, complex journal article, and it will cut it down to a few hundred words. Or simply ask the AI to tell you all about a topic. It will do so with writing that is flat and easy to recognize. It is laughable that a new market for software that identifies AI writing sprang into existence overnight. As if it's hard to spot. The sentence structure is monotonous, with a few equally long sentences per paragraph. Each paragraph a new topic, avoiding building toward anything. No style. It is not even a poor style, as someone might at least say of human writing, like mine. AI writing is not even wrong. Still, brief summaries and answers serve their purpose. The leap from a Google to AI was coming. Why click link after link and scroll through endless text and reports to see if someone has a decent short answer when asking the AI outright will get right to it. The AI will be more succinct, at least, than the proper answer buried in the twenty-two-page version.

And yet, my experience, even with a bit of AI, has shown me – in summary – that the aerospace sector must learn to grapple with the age of AI, whether it likes the writing on the wall or not. Better yet, we may find an indispensable ally in the algorithms.

I'M WITH THE AI, AND I'M HERE TO HELP

The same human who helped create the AI had only one task at this moment, move the stone to its place on the board as the AI instructed. The move would seem to be a bad move, except later when it seemed the AI was playing in a way we humans could learn from. This is a scene from AlphaGo – The Movie, celebrating the humans that created such a powerful artificial intelligence/AI at the same time we feel for the plight of the human opponent. Programming an AI to play Go presents a problem that also occurs in launch and space systems design, more so as we stray far from what's known.

> "The professional commentators almost unanimously said that not a single human player would have chosen move 37."
>
> -ALPHAGO - THE MOVIE

I've written on reusability as key to sustainability, and there being so many design and technology decisions. On top of "what" design decisions to make, say legs and fins vs. wings and things, or a middle path, with even more decisions past those forks in the road, there are more decisions about "how" to organize people to get it all done. There is product and there is process. Human intuition, ingenuity and bursts of inspiration are irreplaceable, yet as our problems get more complex, could we use those same powers to create a helper?

Go has 10 to the power of 170 possible board combinations (more than chess). Back in 2013, finding that even a simple model for a reusable launch vehicle design created a vast number of possible combinations I dived into genetic algorithms (a low kind of AI). The design model I created contains only 97 inputs with 2 to 5 choices each (3^14 unique selections). In NASA-speak we call this a "simple" model – as many years of work as went into it! Even so, the total number of unique design input sets for this model reaches 10 to the power of 47. Not quite Go, but in the neighborhood.

Checking every combination to find the best design would be impossible and irrelevant, as how these unique design choices combine is complex. It's the same as in life, a choice may be great for you here and now, not so much in the future, or vice versa, and to boot it's all limited at the end by a practical matter – your available funds. But as AlphaGo showed, winning the game can be about winning *just enough*. Completely routing the enemy is simply wasteful.

The first thing the AI does is re-orient your reality

You may have told the AI the flavor of each little decision. Then it's the AIs turn to tell you what it all means when put together – and it may be a surprise, like move 37. Normally it's pay me now or pay me later, as seen on the graph on the left below. I MUST spend more to get a future benefit – right, always? In practice it's much more complex, as *what* we want is also affected by *how* we do it. There we locate counter-intuitive design and organizational combinations to spend less up-front *relative to another path*, yet still reduce future costs – as in the graph on the right below.

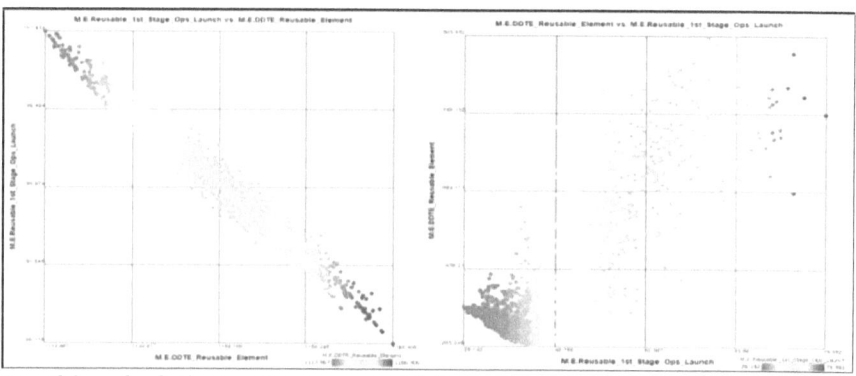

For this calculation, set your very high-performance personal computer in motion before going to bed and hope the answers have spit out by morning. This analysis was created using a genetic algorithm, a poor man's AI. Each single point is a reusable launcher design containing hundreds of smaller, particular design, technology and organizational decisions.

Of course, moving away from engineering circles this is not counter-intuitive at all. NASA found this out when it looked at how its models failed so miserably predicting Falcon 9 costs. We try, we learn, and we discover that sometimes you can have it all. You can spend less, much less even, yet get more - all by turning the knob dramatically on "how" to get "what" you want.

Such an AI approach does not make Starship RUDs any less required. Experimenting leads to learning. And there is no replacement for experimenting with different kinds of real space systems. In the real world, attendance is mandatory. Yet, how does each player decide what it's next moves are in this game, and given all the learning? Human intuition, creativity and inspiration will never be replaced. But, perhaps one day the players designing reusable launch, asteroid miners and refueling stages are just as helped by the AI they created to take all the possibilities and make move 37.

AI, ART, WRITING, OH - AND SPACEPLANES

There are the facts, and there is the story. Both can be true, but one without the other is incomplete, as it is the story that carries meaning. As the AI ChatGPT consumes my social media feeds, it's enough to make a blogger worry. One day soon, will an AI put the words and the melodies together and write the songs - and the stories too? As Liam Neeson said, "what I do have are a very particular set of skills, skills I have acquired over a very long career." This results from experiences, not only with projects but with people. Circumstances, contexts, and people add meaning to my knowledge, thoughts, or misconceptions. We are our stories. Among these experiences is trying to do what an AI might do one day, using massive amounts of information to illuminate the road ahead.

My story today begins back when an AI fit on a floppy disk. For those unfamiliar with the term, these were little plastic disks (I'm showing my age). Think vinyl for your retro record player from the fad resurrecting LPs. (OK, vinyl is not dead, and it's still a thing everyone does with their latest album, from "Best of" oldies to Dua Lipa. But I digress, which is very human. Will an AI ramble one day? Come back to that.) So, these floppies stored data and programs, and this 5+¼ inch floppy disk promised an AI for searching your files. I'm reminded of a lost weekend on this one. Of course, Google eventually created the ultimate AI for search, courtesy of selling ads, as a front for the actual business of trafficking in personal information.

Years later, still on dial-up, I discovered a program that would archive web pages crawling around all by itself. (Digressing again, "dial-up" was when we had a thing called "landlines"…OK, I'll stop here.) Of course, I set my computer to run all weekend - till it froze up nicely. The program couldn't tackle such an endless task, no matter my storage space. Right after, The Internet Archive came along and did the job, operating to this day.

Today, large language models are taking in massive amounts of information indexed and accumulated over decades. Our thoughts are now the bits and bytes these models consume whole. Until recently, chat boxes took about 15 minutes of my time before I moved on. Now I've renamed my ChatGPT shortcut "Time Suck" to remind me to set a time limit. To judge by social media feeds, for many, this has been a thing for much more than a weekend.

It's easy to be alarmed about what may come of all this. If you are an artist, the news before the ChatGPT buzz already sounded like your end-days are near. Multiple AIs have also been digesting vast image libraries and regurgitating rarefied fusions of these as new images when asked to render a kitten joining the styles of Warhol and Mehretu. It's looking bad if you're a web artist hustling for commissions.

Yet perhaps we are getting ahead of ourselves. I had the good fortune to watch 2001: A Space Odyssey and Colossus: The Forbin Project when I was still a child. This was a curious combination, my first exposure to AIs – the first sympathetic if dangerous, the latter unfeeling and cruel. Ten years later came along my favorite evil AI of all time - Skynet. More recently, children are growing up with stories of super-intelligences that put us all in vats (The Matrix), TV series with battling AIs (Person of Interest), or an AI sending people across time (Travelers). "The Director" in Travelers, never referred to as "it," has a grand plan. Choose your favorite dystopian future with an AI making mischief – there's plenty to choose from. In all these stories, what we don't have is a good picture of what an AI might do with the rest of its time.

※ ※ ※

In NASA, we were fond of saying, "all models are wrong, but some are useful." I'm no stranger to models and a poor-engineers version of an AI – genetic algorithms. (These algorithms also take all night to run, now on my laptop, successfully and mostly unattended. See a common theme here?) In the late 1990s, as we looked to design new space launch vehicles, a fundamental problem was everyone's models were asocial. Your model did not "talk" to my model. So much for "One NASA." When we first attacked the problem, not everyone thought it could be fixed anytime soon. I saw plenty of "one day, far, far, away - but not now." The arm waving included showing off your complex model, too complex to connect to anything else without the expert in the loop. Alternately, some argued their models were not complicated enough, being so narrowly crafted. How would anyone, except the maker, know which unique tool to pick from in the garage – or even less, how to use it?

More likely, much of this was about control, the desire not to lose it. If anyone could spin up models across NASA centers and design a spaceplane, what were all the experts to do? My retort was *"build better models" - to create ever better spaceplanes*. This included building new user-friendly interfaces for the non-experts. (Sounds familiar, non-artists and non-writers who are now generating art and essays?) My naivety was apparent. Yet the project carried on just long enough to see the possibilities.

We gathered at Marshall Space Flight Center for an all-day workshop, cranking up our models and writing real-time scripts to connect them. This marked a moment when the end-to-end connection of models created a mega-model. Albeit, this quiet hurrah soon met its nemesis in a combination of sticker shock and firewalls. To boot, the vibe persisted about the tools being misused, with automation leading to incorrect conclusions. True, the technology was not ready, but it didn't help that neither were many of the model owners. Bespoke analysis, experts in the loop, workers unite!

※ ※ ※

Now we know, you can't hold back a good thing, or an AI. I wrote about this in May of 2021 - "I'm with the AI, and I'm here to help" There, I showed the use of a genetic algorithm to make reusable launch vehicle design decisions. Genetic Algorithms are a low-level of AI, rather basic. These programs seek solutions by experimenting with a model at first to see what happens and "seeding" subsequent results with only the better ones from previous generations.

To appreciate the reason we need such techniques, we can count possibilities. Say you have a bunch of design choices to make, perhaps a few hundred. The number of unique combinations of these choices approaches *10 to the power of 47*. Give it a try looking for the best combinations manually - but only for a while. That's till you realize that adjusting one variable to do better near term is causing trouble later, and vice-versa. It's then you call the AI and make peace. You may not understand how it gets its results, but it works. (As the Travelers said, "Trust the Director." Some excellent pieces about the inability to trace an AI's answers, with valid concerns, were recently published in Vice and Noahpinion.

A lot has happened in the world of AI in the last 18 months. How appropriate – the AI field moving so fast it's as if fact is following fiction. Fiction aside, we see AI applications in materials, discovering new invar alloys with ever lower thermal expansion coefficients. Or imagine finding new quantum materials with an AI assist.

For NASA, it's an obvious leap to see how space transportation and space applications could also be helped by an AI. If you have a complex problem where improving in one spot causes problems in another, oh and by the way, you're missing some technology that does not yet exist, you have a perfect application. Here, Northrop Grumman still uses the model-of-models approach we experimented with in the late 1990s, but to optimize the design of a stealth aircraft. If we want to see regular passenger service to low Earth orbit, spaceplanes will fit the bill – except we have only the parts and pieces for figuring these out right now. But we know an AI (and a supercomputer) loves to gobble up such complex problems.

Similarly, as a reason for going to orbit at all, AIs will soon be finding new drugs for medical treatments or predicting drug response. (Think genetic algorithms, pun intended.)

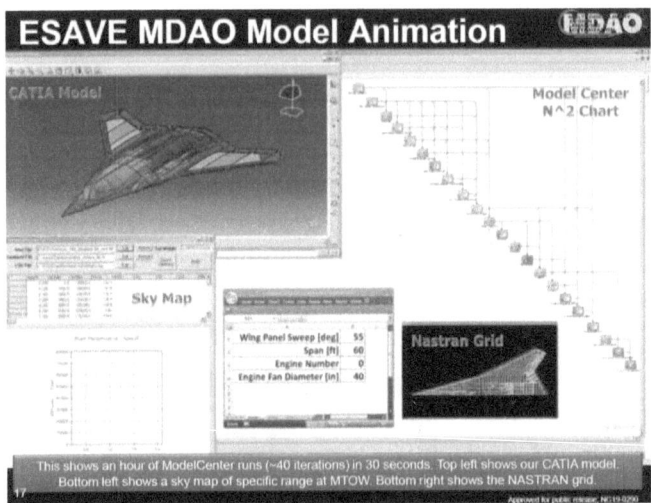

Join up enough models and you get a large model-of-models where you can see the aircraft shape change as algorithms try to find the combinations of inputs that create, on balance, the best results. From "Multidisciplinary Design Analysis & Optimization (MDAO) at Northrop Grumman," Kentaro Sugiyama, July 2020.

Soon we may have an AI that generates cat videos, or better yet kittens. The cats and kittens we see won't be real, but if you can't tell the difference, will you laugh less? Or we will have other possibilities besides unemployed cats. Spaceplanes that breathe in air, more flying engines than aircraft, could use a dose of AI. The scramjet and hypersonic data are out there, and there's no lack of prior work in NASA and the DOD. Among more of my stories would be ones where all the numbers said "go down this path" but agencies had neither the funding nor the interest. Too often, agency leadership questioned the technology and the numbers. Perhaps it's time we trust the AI.

"I DIDN'T UNDERSTAND A WORD YOU SAID"

Recently, the buzz around AI has been about being untraceable, the inability to explain why an AI does what it does. Try and backtrack through an AI's logic step by step, and you are Alice going down the rabbit hole. This is so for those who create the technology, intimate with it to the most minor details. Now imagine everyone else.

This topic of AI and unexplainability already made the rounds in 2019. The baffling moves of an AI playing Go had already drawn attention. Ever felt someone you read about, or someone you knew, was operating on a whole other level? Now imagine an AI on levels you are not aware exist. Recently you will find excellent pieces in the New Yorker, with an analogy to data compression and "loss of fidelity," or over at Noahpinion from the point of view of a choice between predictive power or control.

Yet are we really in new territory, or did we go in a circle back to where we started?

"I didn't understand a word you said." This NASA manager did not mince words.

And here I thought I really took this down to the brass tacks. The analogies, backstory and lots of "like" and metaphors were all in there.

To be fair, my analysis was rather complicated. Also, to be fair, the manager likely aggressively questioned my work due to being more new than complex. But, amusingly, at the time we had only a hint of the complexity to come. Our team knew, too, if Lt. Commander La Forge presented a plan right now for fine-tuning the warp coils on the Enterprise, the reception would be congratulatory. This would be so without the manager even pretending to understand warp coils.

These biases came naturally for an organization narrowly focused on performance, all about the holy trinity of payload, dry mass, and Isp. Shifting the discussion to obsess over ease of reuse inevitably created communication issues. Today, my lucky day, would not be the first or the last time I received this kind of feedback.

❋ ❋ ❋

Let's go back, movie style – "Five Years Earlier." In the beginning, came the spreadsheet. Like tribbles, one became two became four and so on. Sure, the math linked the factors, adjusted, added up, and parsed, and you could follow every single step, and – well, by then, we'd lost the audience. Crash and burn, crater, let's clear the debris and try again. We learned individual calculations, puts and takes, didn't make for a compelling story. The characters were endless, but where was the plot?

Next came *models*. Occasionally we would say "the model" as if it spoke all on its own. Soon after came simulations. "When" now accompanied "what." Space met time, and the evening and the morning were the 3rd day, so we still had a lot of work to do.

Could we think of this as activities, supply chains, or LEGO-like blocks anyone could assemble to explore next-generation reusable launchers? Of course, except now we had to start answering questions about being *verified* and *validated* – which, it turns out, are not the same in this world of models, simulations and inputs and outputs.

For a while, the models and sims spoke with us, like Guru's using new languages and terms. The addition of rigor grew the obtuseness faster than the acceptance. Move quickly, and before you know it, half of the presentation is catching everyone up from when last you met.

"That all sounds good, except this is all so complicated you can *pencil whip* any answer you want, and no one in this room could say otherwise."

(I must admit I had a hand crank pencil sharpener in my office and #2 pencils in a coffee cup. For most of my career, this anachronism remained nearby. Another confession, I still have a pencil sharpener, now in my garage.)

When spitting out answers favoring one design over another, the headwinds are not only about rigor or traceability. *Complexity*, and worse, a flood of information expressed in a new language, will clearly hobble the adoption of new ways of attacking problems. Still, we had a saying among us – if you are taken aback by harsh critique, you must find another line of business. I was not taken aback. Try again, including new ways to communicate, share, listen, and improve.

For context, though, imagine you were just in another meeting where it is said, clear as gospel, that if something weighs less, it will cost less. This is followed by the mass reduction from the Shuttle upgrade costing much more than the considerable expense initially advertised. Oh, and even before accomplished, seeing it will cost more per flight. Mumble, mumble, then many asterisks are placed on the notion "less mass equals less cost" (a thought which, for the better, eventually ceased to be a going concern.)

What does not kill you… well, you get the picture. The tough questioning improved all our models and simulations, and one day we are running "design of experiments." Soon after, we realize we might as well go across the entire life cycle of technology, not just the part on the tail-end at Kennedy Space Center. Even better, the models now have "Reverse" on the gear shift, not just "Drive." Instead of entering a design and getting answers, we would enter the answers we want and ask the machine to give us the designs. The sense of predicting a proper investment direction became more robust, like the hurricane tracks for different weather models all converging on the same landfall. The downside was an increase in complexity that slowed down further work. We picked all the low-hanging fruit and now came the hard part.

❋ ❋ ❋

Needless to say, we also received plenty of encouragement, for which I am thankful. Intuitively, most complex problems have complex solutions that do not make for effortless storytelling. Our sponsors saw this, and with them, our work did not need a hard sell. The best vote of confidence came from projects with real hardware asking us to join their team, even if most of that hardware never flew. (Remember the part about fixations on performance.) Nonetheless, enough audiences saw how answers with a proper sense of direction sufficed regarding *trust*.

For AI, as in many fields, the matter of being traceable versus trusted may prove similar. If you adopt some advice, you will test it in the real world, and if the advice proves correct, you will say, "Success," and move on. Even if you believe the answers came from something akin to philosopher John Searle's "Chinese Box," a non-sentient, merely superbly complex program, you will feel comfortable with the answer. Trust will increase even as traceability declines. If you create your AI's suggested alloy and get a leap in strength, you could be more than fine not knowing how the machine got inspired. You should be excited, not worried.

It's becoming common in aerospace to hear about

optimizing for manufacturing, for low cost up-front, *and* once operational. The complex push and pull between the decisions we face in complex aerospace systems could be simplified by narrowing your focus to only performance. All other factors become results, revealing themselves over time. Costs? Just wait for the bill. Schedule? Wait. To. Finish. Reliability? Safety? You will discover the result there too. Alternately, we can embrace complexity to favor answers and success of a sort we would never see if fixated on narrow, simple questions.

Of course, there is the well-deserved negative critique of inscrutable AIs when they reinforce biases and errors, say regurgitating misinformation from the internet. But that too is old news, since the days we said "garbage in, garbage out." Down other avenues there is so much promise for tackling important problems while respecting their complexity. Going full circle, it's too easy to imagine the future.

The year 2033. She thought her presentation was going well. "Enjoy the moments like this," she said to herself, a chance to share and to learn. Only a few minor comments came here and there from the otherwise quiet audience. Sure, she knew the work was bug-ridden and expensive, with many caveats and limitations. It would take much more work to be truly useful. A perfect time to say, "all models are wrong, but some are useful." Better yet, best not to talk about the bugs and gaps. Those holes didn't matter. Better not to get into explaining Shor's algorithm either, that didn't go too well last time. What matters is using the new quantum frame and our AIs together, already pointing development in wildly creative directions. We are generating new questions, leaps beyond what the company AI did on its own giving mere answers, so limited by what it absorbed, even if that was nearly everything. The really wild answers were right there, buried in the ground states of the new questions – *if we phrased our questions for a quantum computer, that is.*

Wrapping up. The complications could be overwhelming, but she knew years from now, this line of attack would be the norm. Now to see if the audience embraced the move beyond that old AI. Last chart. Turn to the audience, "Any questions?"

TRUST THE DIRECTOR

I've been busy running experiments, well, the algorithms have been busy that is, once I set them running. Years ago, I wrote about using a genetic algorithm to evolve reusable launch vehicle designs, including the best way to pull it off. I called this "what" meets "how." So, this wasn't only about a power supply, a new technology, or the features of your rocket engine. This included how lean your organization ran, how it worked with suppliers, and how you organized. Typically, no one wants to run an experiment and change more than one variable at a time. I thought – why not? Better yet, let's change them all – dozens - simultaneously.

"Trust the director" was one of the many memorable phrases in the Netflix series "The Travelers." The premise was time travelers from the usual dystopian future travel back in time to make the future less of a dreary dystopia. A novel element is they get orders from an AI, the "director." They constantly remind themselves to follow their orders, trusting that the quantum AI from the future knows all and sees all possibilities. This proves easier in theory than in practice. It's easy to lose faith in what we don't understand (with the obvious metaphors to religion and deities.) Not everyone is on board with taking orders from a machine. Sound familiar?

To judge from the news, the battles are raging between the travelers who would say "trust the director" and those who could care less about the correct answer, favoring human choice. Choices –a word I use throughout the paper I just uploaded to academia.edu as a draft. I fall on the side of the director, to say the least, letting the machine do its thing for the simple reason the possibilities are beyond me. A human can set up a game, some rules, and competing objectives (the tricky part), but it takes an AI to explore every possibility. At a time in NASA, we also challenged the directors – the human ones - to make choices and explore possibilities.

Even before the loss of Columbia, NASA leadership was fond of reminding anyone providing advice that it was all merely *for consideration*. Costs, schedule, risk, and reliability were all vague possibilities, whereas doing what was appropriated was real. For a time, you could catch a manager (likely recently returned from training) saying they only focused on what they could control. Viola! How effective. Pick your battles and all that. This sounded great in theory, but in practice, what a manager said they controlled approached zero once a project was approved. The ship was out to sea, the course was set, and even the highest level of leadership shared the wisdom of realizing your job was to keep the engine running. The analysts and advisors reveling in "choice" were, at best naïve. At worse, they were saboteurs trying to steer westerly instead of north, staging a mutiny, starting with taking control of the bridge and the engine room. (Next time, use an AI to figure out how to take both, simultaneously.) Wisdom was in seeing choice as an illusion. If you realize this, you might become a "director," too.

AN ENGINEER'S JOURNEY IN NASA

I am old enough to remember when professors thought spreadsheets were the spawn of satan. We will raise a generation of engineers that can't do math in their heads. But to feel really ancient, I remember handing my calculator to a professor during a test as she walked the rows of desks checking to make sure no one had a "programmable" calculator – the latest thing. (Some did graphs too!) So I am likely not the best, unbiased source for which side to choose in this debate. In my experience, machines were always valuable tools, allowing humans to focus on understanding rather than mind-numbing number crunching. Oddly, in some instances, we don't think twice about trusting AI-ish algorithms that hardly anyone understands. The Traveling Salesman problem is one of these, finding the shortest route for hitting every location on a list. FedEx anyone? Trust lags in other walks of life, and we say, "Not so fast."

What I do know, though I wish I had seen this ten years ago when I first worked on this with the Department of Defense, is when a machine can give an "Aha" moment that otherwise avoids detection. Complex problems are often overcome by sheer firepower and complex solutions – running for days on the laptop. I added that "Aha" moment to the work, which has evolved over the years. Another discovery? I will assume we can make choices. Many. There is more latitude to steer the ship than many believe. And I will trust the director – the AI version - for now.

EPILOGUE

Parallels

Time is supposed to be the way the universe keeps everything from happening all at once.

It doesn't always work this way.

My wife and I were in London in 2016, and we made a point to see The Fighting Temeraire by Joseph Mallord William Turner. London has endless museums. So, we were taking it all in small doses, searching out specific works. This was one. I could feel time passing in The Fighting Temeraire, slowly, then quickly.

The Fighting Temeraire by Joseph Mallord William Turner.

After many battles, the Temeraire is towed to the breakers, having been sold for scrap. A steamship, a lowly tug, is pulling it to its eventual resting place. Technology has moved along, sails soon enough a thing of the past.

Past scenes come to mind after staring at The Fighting

Temeraire for a while, soaking it in. I had walked under and crawled around in the Space Shuttle Atlantis once upon a time. Then, time moved slowly. But as happens when endings approach, time suddenly feels sped up. That was the case as I stood by the roadside in 2012. Now Atlantis was being towed, victory parade and all, to be prepared for the museum at Kennedy Space Center.

The following year, I visited the museum where Atlantis was now mounted on steel beams. The tour guide's voice was a distant hum of inane tidbits. Was it the usual about the seconds it would take for the Shuttle's engines to empty a backyard pool? I could have been in London years later, hearing how it took 5,000 oak trees to build a ship like the Temeraire. Instead, here was Atlantis, frozen forever in a semblance of motion.

Atlantis, Florida. Years later, The Fighting Temeraire. Soon after, a Falcon 9 booster on its drone ship coming in to the port at Cape Canaveral. And I can't help but see it all blur over, as one. Turn the page.

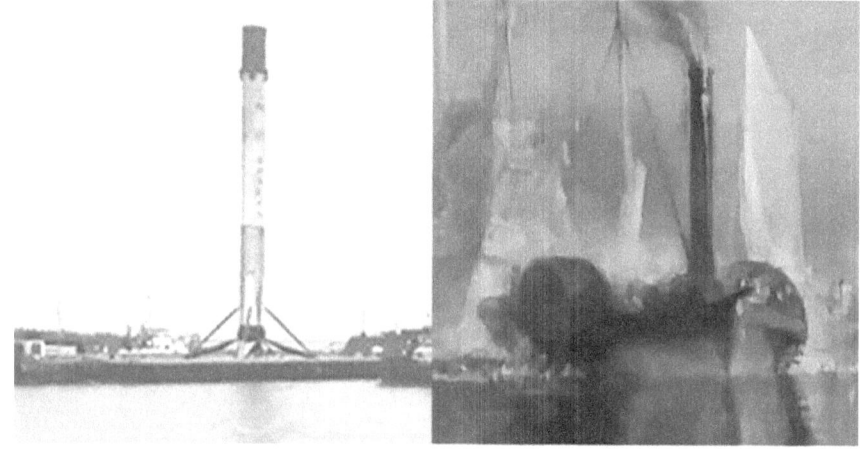

AFTERWORD

I learned a scary thing while blogging. After a few years, it has become rather effortless for me to write. That does not qualify me as a writer. Someone who has strengthened their muscles with weight lifting is not automatically qualified to paint a mural at the local park. Fortunately, as our aerospace world sees rapid change, though some want faster, I am happy to say – at least I will never lack for topics. And it appears I do not lack for words. However, I have not touched upon many topics and cannot share others. Anyone who has ever worked with agencies outside NASA knows this.

In my last few years at NASA, I spent more time supporting a DARPA project, the eXperimental Spaceplane Program, than on NASA matters. This is only one example of how our industry hides invaluable knowledge that should be shared. This is for valid reasons, but to poor effect. So, grab a pencil and paper, your keyboard or journal, or feed a typewriter near a hundred years old like the one next to me. And at that gathering, talk and share. Your experiences matter. Never mind if you take a wandering journey away from the focus topic. Talk about NASA, DOD, start-ups, and companies and what we can or should do next. Your knowledge must not be lost. What have you learned? We might still have much to be surprised about in our lifetime, more likely than not.

ACKNOWLEDGEMENT

I am grateful to my early supervisors who supported the back and forth between my daily duties, preparing and launching, and the work of what would follow. The latter must have seemed a distraction at the time. My thanks, in particular, go out to Russel Rhodes, a lifelong mentor who passed along a wealth of experience from his time before and during the Apollo program, as well as in the development of the Space Shuttle. My thanks also to Carey McCleskey, alongside in the many projects, technology, and tasks, clear of vision and purpose. We were often called "The Three Musketeers" – working together to take the lessons of the Shuttle era forward.

My thanks to sponsors who cleared a path when it appeared others lacked much interest in looking too far ahead. This includes Jay Falker and Charles Miller at NASA HQ and numerous personnel in the Defense Department, Air Force Research Lab, and the Defense Advanced Research Projects Agency. As I can not name everyone, I would like to acknowledge all those who supported my work, even when the answers were not comforting or reinforced the party line.

www.ingramcontent.com/pod-product-compliance
Lightning Source LLC
Chambersburg PA
CBHW031603210526
45464CB00004B/1406